Other books by Tremaine du Preez:

Think Smart, Work Smarter
A Practical Guide to Making Better Decisions

Raising Thinkers
Preparing Your Child for the Journey of Life

DECIDE

the art and
science of
choosing
wisely

Tremaine du Preez

Marshall Cavendish
Business

© 2020 Marshall Cavendish International (Asia) Private Limited
Text © Tremaine du Preez

Published by Marshall Cavendish Business
An imprint of Marshall Cavendish International

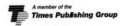

All rights reserved

Other Marshall Cavendish Offices:
Marshall Cavendish Corporation, 99 White Plains Road, Tarrytown NY 10591-9001, USA
• Marshall Cavendish International (Thailand) Co Ltd, 253 Asoke, 12th Flr, Sukhumvit 21 Road, Klongtoey Nua, Wattana, Bangkok 10110, Thailand • Marshall Cavendish (Malaysia) Sdn Bhd, Times Subang, Lot 46, Subang Hi-Tech Industrial Park, Batu Tiga, 40000 Shah Alam, Selangor Darul Ehsan, Malaysia.

Marshall Cavendish is a registered trademark of Times Publishing Limited

National Library Board, Singapore Cataloguing in Publication Data

Name: Du Preez, Tremaine, 1978-
Title: Decide : the art and science of choosing wisely / Tremaine du Preez.
Description: Singapore : Marshall Cavendish, [2020].
Identifier(s): OCN 1129112604 | ISBN 978-981-48-4159-7 (paperback)
Subject(s): LCSH: Decision making.
Classification: DDC 153.83--dc23

Printed in Singapore

This 40,000-word book will take about
two-and-a-half hours to read.

That's less time than an adult education course and
slightly more than a superhero movie.

The payoff from this time invested is unlimited.

CONTENTS

Preface • 9

Part 1

1. Your choice • 17

2. What is a good decision? • 19

3. Must a good decision be rational? • 23

4. Can we avoid thinking clouded by emotions? • 31

5. Must good decision making align with our values? • 37

Part 2

6. Your current decision-making process • 47

7. Process vs goal orientation in decision making • 51

8. Decision rights • 61

9. The meta-decision • 67

10. The business of bias • 79

Part 3

11. Risk and uncertainty • 111

12. We are all Brandon Mayfield (Case Study) • 124

13. Unconscious processes • 130

14. Gender differences in decision making • 145

15. Stress is an emotion, too • 149

16. The pain and power of alternative opinions • 155

Part 4

17. Your enhanced decision-making process • 164

Over to you • 174

Author's Note • 175

References • 176

About the Author • 183

PREFACE

Does the world need another book on decision making?
Who knows more about decision making:
practitioners or academics? Why theoretical case studies
won't cut it. Choosing your own learning journey.

There is no shortage of books on decision making, which is why I wrote this one.

Let me explain.

If, like me, you've read your fair share of books on this topic, then you'll know that there is a frustrating polarity between the pages. Academic books written by researchers bring us the theory and science of good decision making. Books written by practitioners who work with decision makers and have years of corporate experience in the application of choice strategies, reflect the art. Academics tend to turn their research findings into well-written, well-referenced offerings, but are mostly compelled to focus on a narrow area of research, usually their own.

Traditionally, academic research in the area of decision making has been carried out on students and other readily available participant groups. This leaves it up to the author to illustrate how their findings would perform with executives who, unlike students in a controlled scenario, make decisions under conditions of stifling uncertainty, extreme stress and faced with outcomes that could significantly impact themselves and others. Making the leap from academic experiments to corporate decision making requires explanatory narratives, fictional characters and life-like case studies. These make for more interesting reading—but, personally, I find that I am quite capable of applying research findings to my own situation and don't need superfluous pages of invented scenarios.

At the other end of the spectrum, practitioners bring essential insights into the real, messy world of decision making, but often lack the academic grounding that would give their ideas depth and breadth. Occasionally a book is published that spans the practical and the theoretical, like Thaler and Sunstein's *Nudge*.[1] Firmly rooted in practice, it shows how behavioural insights gained in a lab can be applied to influence behaviour towards practical outcomes. Another is Sabrina Hatton Cohen's *Heat of the Moment*, which is part memoir of her life as a firefighter and senior incident commander, part exploration of her PhD thesis on decision making during critical incidents.

I'll confess: I first learned about decision making through the work of management gurus, who earned fame through eloquent narratives and easy to apply step-by-step formulas wrapped up in memorable mnemonics, optimised for sharing on a Twitter feed. Then I stepped into the academic world and was expected to justify the professional, practical and tacit knowledge that I brought to my research after 20 years at the coalface of industry. I trudged through academic libraries to verify the claims I had been peddling for years, only to find a great number of them to be factually incorrect, outdated or so oversimplified they were misleading. This was a painful and humbling lesson, but one I am very grateful to have experienced. I won't name and shame, but many of these ideas will be put to the test here in a stand against alternative facts and pop psychology.

So is this book art or science? Well, I am first and foremost a practitioner in organisational decision making. This puts me in the interesting position of having access to a library of real-world challenges and decisions made. Examples used in this book are true but anonymised so as not to get into trouble with my clients, colleagues and friends. However, I prefer to stick to the point, so won't subject you to unnecessary narratives if a simple explanation will do.

Having researched a doctorate degree in decision science, I can't overemphasise the importance of a sound knowledge base from which to draw conclusions and, if I dare, give advice. The work I have done in this field, along with the time very generously given by organisations and decision makers, have transformed both my understanding and application of the science and the practice of good decision making. However, this book isn't only about my current research[2] and so can be as wide-ranging as needed to answer the question: what is a good decision?

I get asked this multiple times a week in both formal and casual conversations. To answer it in a 30-second elevator pitch would be a flippant oversimplification. Furthermore, to give the impression that there is a definitive answer would be misleading. The real answer is: it depends.

Depends on what? is the question this book attempts to answer. It unites my professional and academic practice with the work of leading authors and researchers to offer a theoretically-grounded, yet battle-tested, practical answer. This will ultimately help you improve the quality of your decision making.

Many practical elements originate from my work as a consultant specialising in organisational decision making. This includes: working with organisations to understand their overall decision-making personality; how organisational, idiosyncratic and market influences shape it; and how to improve the quality of decision making in line with organisational objectives. Yes, it is a big topic and not one that I ever want to lead with at a cocktail party. It usually provokes a sympathetic smile followed by a change of subject.

The following is a brief overview of the book, allowing you to focus on your areas of interest:

Part 1 begins with a theoretical history of rationality and good decision making. It explores the research underlying the most popular responses to the question, 'What makes a good decision?', including addressing what we know about what works and what doesn't. I would highly recommend skipping Part 1 if you are a practitioner, unless you can't sleep, and theory makes you sleepy. Don't say you haven't been warned!

Part 2 examines your current decision-making processes. This self-diagnosis is an important step on the road to improved decision outcomes. I recommend you don't skip it. Decision making is intensely personal, and this book will not provide you with a five-step plan to making great decisions (I'm not a management guru, after all). My suggestions and insights are offered to augment your current strategy so that it works for you and complements the way you already process information and come to conclusions, especially when under pressure.

Part 2 also introduces a decision-making process that brings decision-making best practices together, exploring each in turn—the research, the practice and only those examples that are absolutely necessary to provoke thought. If you are already committed to a particular decision-making process (chapter 7), then start at chapter 8. The latter introduces the first few steps in a good decision-making process, namely, understanding and allocating decision rights, crafting a meta-decision and exploring the power of debiasing strategies through examples.

Part 3 continues exploring the elements of a best practice decision-making process through universal risk assessment strategies and the role of emotions and gender in our thinking and risk perception, as well as providing a case study for you to try out your new skills in evaluating an FBI fiasco.

Part 4 wraps up with some basic decision-making hygiene, chapter summaries and a cheat sheet to help you put what you've learned into action in your everyday decision making. If you are a student of mine or have attended a corporate programme run by my consultancy, DECIDE, then jump to Part 4, you'll know immediately what's going on. Of course, I will offer plenty of other resources, authors, books and journals for those wanting to dive deeper into specific areas.

Ready?

PART 1

1

YOUR CHOICE

> "God is, or He is not. But to which side shall we incline? Reason can decide nothing here … A game is being played at the extremity of this infinite distance where heads or tails will turn up. What will you wager?"
> —Pascal's Wager, *Pensées*[3]

Pascal proposed that to exist is to participate in the ultimate wager—to choose between two uncertainties that reason cannot illuminate. The first is that there is a God and an afterlife of peace and prosperity will follow a life of sacrifice and devotion. The other is that there is no God and piety earns no otherworldly rewards.

What do you choose to believe? How do you decide? Perhaps your risk profile sways your thinking—do you enjoy a gamble or prefer a safe bet? Perhaps your time horizon informs your choice—do you prefer to make the most of the here and now and leave the future to take care of itself?

Or perhaps you wondered about the odds of there being a God omnipotent and benevolent enough to create an afterlife sanctuary to incentivise devotion? Life, it seems, is a game of chance. Such games have attracted mathematicians as far back as the 16th century with attempts to analyse outcomes through probabilities and how to cheat at them convincingly. The mathematical treatment of Pascal's Wager was the first recorded use of formal decision theory in Western philosophy and ground-breaking in its contribution to the brand new field of probability theory. It subsequently made possible the ever-familiar bell curve, regression towards the mean, subjective probabilities, utility maximisation, formal risk analysis and many of the other theories and tools that have filled the canon of rational choice over time.

Whether you decided to believe in a God or not, do you believe you've made a good decision? What criteria can you use to judge this decision? This leads us to the next question for you to answer:

In your opinion, what is a good decision?

2

WHAT IS A GOOD DECISION?

A crowdsourced answer to this question, busting some myths about what a good decision is and why we can't judge a decision by its outcome.

What did you decide? Does your answer include one or more of the following?

A good decision:

1. achieves its objectives
2. logically considers all the options at hand
3. avoids thinking clouded by emotions
4. aligns to an organisation's or individual's goals and values
5. avoids regret

If it includes any of these things, you're in good company. Since 2008, I have posed this question to thousands of people from over 21 countries across five continents, and the answers I receive have been remarkably similar, as summarised above. Yet over more than a decade of researching and working to improve individual and organisation decision making, I have learned that only one of these answers is truly valid across different circumstances and problem domains. Any idea which one?

Let's start at the top. Surely a good decision achieves its objectives? This is the most popular response received, from CEOs of multinational companies and senior leaders to secretaries and support staff. If only one of these is valid, this must be it?

Let's apply a bit of critical thinking here—something we'll be doing a lot of together. In deciding on the objective that the decision is to achieve, how can one be sure that it is the best possible objective? From whose perspective? Over what time period? Who is the best person, or persons, to judge that? Did the decision maker even solve the correct problem? What if forces completely beyond the decision maker's control—such as a trade war or some environmental, political or corporate shenanigans— batted their efforts off course and resulted in the chosen course of action not achieving its objectives? Does that mean she or he made a bad decision?

> **"** You can't tell by the outcome whether
> you made a good decision.
> It's just a logical mistake to say, 'I got a
> good outcome, I must have made a good decision.'
> And yet that's what everybody thinks. **"**
> —Ronald Howard, Professor of Management Science, Stanford University

So, can we judge a decision by whether it meets its objectives or not? Probably about as much as we can judge a Netflix series by its title artwork.

Although we can't ignore the fact that, as a professional decision maker*, you *are* judged on the outcomes of your choices—your income or bonus reflects how well you've attained your objectives or targets and your professional reputation builds on achievements made, not the process used to achieve them. This is pretty much the status quo today.

However, luck or a poor process will not result in consistently good decision outcomes over time. In chapter 2 we will explore the complement to an objective or outcome-focused approach to decision making—namely, process orientation. A good decision is underpinned by a good process. And before you say that sometimes you make great decisions on the spur of the moment without a process, we'll also explore the mechanics that underlie good rapid decision making as well as gut feel.

What about number 2 in our list of answers above? That a good decision *logically considers or weighs all the options at hand?* What do you think? Is this even possible?

Sadly, one cannot weigh *all* the options available but only the options that one is aware of. So we can cancel that answer immediately. But what about logic? It would seem obvious that it plays a starring role in good decision making. From the Greek *logos* meaning 'reason', logic describes 'reasoning conducted or assessed according to strict principles of validity'.[4] Logic is foundational to rationality, which is 'the quality of being endowed

---

* Have you ever considered yourself a professional decision maker? If you are required to make decisions on behalf of any kind of organisation, then you are one.

with the capacity to reason' logically. Apparently, this is 'a trait that distinguishes man[kind] from animals'.[5]

Mankind's capacity to employ rationality and hence, logic, in reasoning was the first, and is still the most endearing, perceived measure of good decision making. But there's a little more to it than the *Oxford Dictionary* lets on.

3

MUST A GOOD DECISION
BE RATIONAL?

Gambling, followed by historical decision theory,
contemporary decision theory and
I'm-never-going-to-remember-this theory. Proof that both
decision theory and our brains aren't as efficient as we think.
A reminder that this part of the book is only for those few
readers who enjoy theory, or those with insomnia.

Let's play a game of chance, a coin toss that will cost you $20 to
participate in. In this game, you are asked to choose between two
gambles:

1. the opportunity to win $400 with a 20 per cent probability, or
   nothing; or
2. the opportunity to win $80 with a 40 per cent probability, or
   nothing.

If you're quick, you've already worked out that the expected
payoff [6] from option 1 is $80 and from option 2 is $32.

Which gamble do you prefer?

According to traditional decision theory, there is only one rational answer here, which is to choose the gamble with the highest expected or probability-weighted payoff—in this case option 1. If you chose option 1 and would do so under all and any circumstances then congrats, you just passed the oldest test of rationality. Option 2 is wholly illogical. But what if you chose option 2? Perhaps a surer thing with a lower payoff is more your speed? What if you decided not to partake? $20 in hand will buy lunch and a train ticket home, gambling this away for some expected payoff may not seem particularly rational to many of us.

As far back as the early 18th century, it was evident that maximising the payoff wasn't on everyone's agenda. Swiss mathematician Daniel Bernoulli (1700–82)[7] explained our seemingly odd choices by introducing a theoretical pauper who was fortunate enough to chance upon a lottery ticket offering him an equal opportunity to win a large sum of money (say $10,000)[8] or nothing at all. Clearly, he has nothing to lose by taking the gamble with a probability[9] weighted value of $5,000. Yet, might he not be better off selling that ticket for less than $5,000? His situation means that he would value even $1,000 in hand more than a 50 per cent opportunity to win $10,000. The utility, or subjective value, that he gets from the former is higher than the latter. Following this, considering a decision not in terms of monetary outcomes but rather in terms of maximising one's utility from the money earned was no longer illogical.

Two hundred years later, decision theorists[10] refined Bernoulli's expected utility theory, this time proclaiming mathematically that to be rational, a choice should comply with three decision behaviours:

1. We must be able to rank *all* possible outcomes to a choice in terms of our preference and stick to them. So, if we prefer red jerseys to blue and yellow ones and blue to yellow, we wouldn't choose the yellow jersey over the red one (even if yellow was the go-to colour of the season).

2. If pink and orange jerseys were added into the mix we would still have to prefer red to blue to yellow, and couldn't take a chance on blue if red was available.

3. If 10 Christmas-themed jerseys were on sale in our size, then it would be rational to choose the one with the most red to maximise our pleasure or utility from the purchase.

Sounds pretty rational, albeit a tad boring and far too disciplined to explain how we actually make decisions. Allowing for the fact that you may have a friend who wears only black turtlenecks or wouldn't think of eating anything other than tacos on a Tuesday, such discipline is not the norm.

These principles can be recognised as the 'strict principles of validity' (*Oxford Dictionary*, 2018), or logic, against which rationality is secured. Mathematical proofs of expected utility theory show that a rational decision maker will always select the option that maximised their expected gain (utility or pleasure) for a particular level of risk.

Unfortunately, this doesn't describe us on even our most rational days.

For this theory to be useful, the decision maker has to know upfront the probabilities of all outcomes occurring and how they feel about each outcome. For example, if you are playing a board game, then the probabilities of a fair dice landing on each number from 1 to 6 are known. However, in everyday decision making we don't always have the luxury of a full suite of known probabilities, or even of knowing how we would feel about them at some point in the future.

Despite this obvious flaw, this technical treatment of rationality dominated choice behaviour in both theory and practice for almost half a century, largely because mathematicians and economists had the field all to themselves. They had decided that it should be used as a guide to make rational choices, as a benchmark to judge the logic employed in a decision and, furthermore, to describe how people actually made choices.

And they could have gotten away with it if they hadn't used it for that last point, i.e., to describe how we actually make decisions. Describing human behaviour with all its contradictions and inconsistencies is not something that mathematicians are well equipped to do. This is the purview of psychologists and by the 1970s the time for them to weigh in on the decision-making debate had come. After all, shouldn't decision making in theory reflect decision making in practice? And so, for the first time, the theory of decision making brought psychologists, economists and mathematicians together in the same playpen, and they've been playing together quite nicely since then.

Far from being a mistake to be buried in online libraries, the early ideas in mathematical decision making have fundamentally shaped contemporary decision-making research, as its inability to model actual choice behaviour became the springboard for a proliferation of activity in decision making, including the obvious question: *If we aren't rational, then what are we?*

> "Saying that we are irrational
> is like saying we don't have fur."
> —Daniel Kahneman[11]

Many theories popped up to answer this question, but I don't want to get too side-tracked here—after all, we are meant to be exploring if good decision making *logically or rationally considers all the options at hand*. More modern definitions abound that take account of our limited capacity for rationality. Among them, rather than labelling us as categorically irrational, we are seen as being subjected to 'bounded rationality' or 'satisficing'.[12]

Apart from the constraints of limited processing power and incomplete information (even if we had full information, we could not completely process it) additional limits are always present as we choose between courses of action—such as limits to money, time, capability, emotional capacity, imagination and others. In fact, economic models that include bounds on rationality have notably more success in describing actual economic behaviour, partly because bounded rationality adheres to a fundamental economic principle of scarcity of resources—except in this case the scarce resource is human cognition![13]

Does this mean that logic and rationality cannot be criteria with which to judge a decision? That good decision making can't include *logically considering all the options at hand* as suggested in my crowdsourced hypothesis? It would be very tempting at this stage to answer 'yes' and exclude logic from considerations of good decision making. But perhaps, like me, you also feel a little uncomfortable with that? If we can't really lay claim to logical considerations of the options at hand, why is it cited by decision makers as so important in decision quality? Surely there must exist some form of logic or rationality that we are able to rely on in our thinking? Despite the evidence against us, if the latter isn't available to employ in our decision making, then how can we trust our decisions or evaluate those of others? How can we judge decision quality?

What psychologists and behavioural economists have discovered is that decision makers are prone to violate the rules of logic and rationality in systematic ways. The research of psychologists and Nobel laureates Daniel Kahneman, Amos Tversky and others opened up the possibility that decision making is systemically irrational at an idiosyncratic level. Even today, the field of behavioural economics continues to remind us that decision making is not driven by numbers, spreadsheets, facts, models or computer algorithms, but by something that is infinitely harder to identify, quantify and label and hence infinitely more difficult to improve upon—the physical and psychological assets of an individual decision maker. We will delve further into these limitations as we build a custom decision-making model for you in Part 2.

How the perfectly irrational make good decisions

What would you consider rational or logical in decision making?

When asked to explain or defend a particular decision, the starting point is usually to explain the context in which the choice occurred. This background information frames our thinking about the problem and our pursuit of options. You might be perfectly happy with your salary, only to discover that a co-worker with the same responsibilities earns 20 per cent more than you do. This might incite you to leave the company, demand a salary adjustment or grudgingly accept this position, because you are a single parent and the economic outlook isn't great at the moment.

One of the first insights that differentiated behavioural economics from rational choice theory was that we consider our options in terms of relative positioning. Relative to other options, to the past, to expectations of the future and our own context, risk tolerance and time horizon. Decision making really is quite messy, but the good news is that we aren't strictly irrational, and some authors have tried to explain our less-than-perfect relationship with rationality.

In the intuitively appealing theory of ecological rationality,[14] we learn that oftentimes violating the principles of rationality is the most rational foundation of a sound decision. This is because the quality of our decisions depends on both our internal criteria and the nature and context of the problem at hand—something we will explore later in problem domains. Yet I wouldn't want to throw out some tried and tested measures of rationality, such as understanding the possibilities and probabilities of outcomes occurring. We still need a set of facts and sensible possibilities to work with. In the salary example above, even if we decide to take the traditionally irrational route of accepting 20 per cent less than a co-worker, we should first establish the possible outcomes to our range of options and the probabilities of those outcomes occurring.

An alternative and more realistic definition[15] of rationality is one that incorporates an understanding of how our physical and psychological assets, their strengths and limitations, affect our decision making coupled with more traditional decision-making strategies, such as consideration of the possible consequences of a choice as well as the possibilities of those consequences occurring.

The introduction of psychological assets has been an important addition to understanding judgement and decision making in practice and narrowing the gap between normative (what we should do) and descriptive (what we actually do) theories of decision making. Specifically, we need to consider the effect that unconscious processes (as part of our psychological assets) have on our perceptions of risk. Does this mean that to be rational we must understand the effect that our physical and psychological assets have on our decision making?

Yes, but it isn't as complicated as it sounds and leads us onto the next criteria of good decision making that we're going to challenge: that good decision making *avoids thinking clouded by emotions*.

How do you feel about that statement?

4

CAN WE AVOID
THINKING CLOUDED
BY EMOTIONS?

What you already know about emotions plus quite a bit
you may not know about emotions. How collective
emotions become self-fulfilling and the relationship
between specific emotions and risk taking.

This is such an important aspect of how we decide, supported
by a vast body of relatively recent knowledge bringing insights
and ideas that are essential to good judgement, that a chapter will
follow on the role of emotions in decision making. For now, let's
just set the groundwork and decide if we can make decisions not
clouded by emotions and examine if this is even helpful in good
decision making.

I undertook a pilot study at the start of my doctoral research. It
had two parts: the first was about rationality and making unbiased
and non-emotive decisions. In this first section I asked 53 senior
decision makers if they made important decisions in a logical
and rational way (I didn't offer a particular definition of either).
Seventy per cent of them answered, yes, they did—of course. Part
2 of the survey focused on emotions and the role of intuition and

gut feel in judgement. Here, 80 per cent of the same respondents felt that emotions were valuable and regularly incorporated them into their decision-making process. Intuitively these respondents were acknowledging their use of a more contemporary form of rationality, complemented by emotion and intuition.

We should probably start by getting clear on what an emotion is. If I asked, you might give me everyday examples—happiness, sadness, anger, frustration etc. But these are merely the expression of the emotion—how they make us *feel*. Emotions are the physical processes underlying these feelings.[16] For example, to feel an emotion, you would have to experience something that gives rise to thoughts about the experience. These thoughts would incorporate long-term memories of related experiences (no matter how remote) that would be combined with chemical information from other areas of your brain and body. All this information would be gathered together really quickly and unconsciously in your working memory. Only after it had bubbled to the surface and caught your attention would you experience this as a feeling. We can think ourselves into a state of fear (top-down emotion), or fear the knife in an assailant's hand because of what we already know about the dangers of a dodgy-looking chap brandishing a knife in front of us.

Defensive emotions such as fear are generally bottom-up, while social emotions such as compassion and shame are mostly top-down and arise from our conscious appraisal of a situation.[17] All these emotions are accompanied by a change in chemical composition in our body, each one with a specific chemical signature. We know these chemicals as hormones. This is where it gets interesting, because researchers can now map the effect on our risk perception, and hence decision making, of the various chemicals that are generated by different emotional states.

Far from being able to remove the emotional component from our reasoning, emotions are assumed to be the *dominant driver of most meaningful decisions*[18] that we make. Ultimately, they guide our attempts to increase well-being and positive feelings while—you guessed it—encouraging decisions that avoid negative experiences.[19] Emotions are perhaps one of our most sophisticated survival techniques and decision-making tools. They influence both what we think and how we think.

Once triggered, each emotion provides a set of instructions via chemicals (aka hormones) to the body to prompt certain actions. These can vary in intensity and quality depending on the situation, but all save mental processing by setting in motion a tried and tested set of responses covering our physiology, behaviour, instructions on how to process incoming information[20] and how to communicate—none of which we have to think about.

Are we at the mercy of our hormones?

That sounds pretty powerful, but is it even possible? What I've described seems to be a master-slave relationship, yet after 200,000 years of evolution it seems unlikely that we are still slaves to our passions. 'Slave' is certainly too strong a word, but think of the first time you fell in love. Your body felt different, your outlook on life was rosier, challenges weren't insurmountable—you might even have felt emboldened to take on a few riskier projects.[21] Such periods of temporary positive arousal will promote increased risk taking,[22] whereas fear does the opposite.[23] Investors faced with rapid losses from a stock-market crash tend to enter into a mental slump that dampens risk taking. The extent to which this happens can be measured by the levels of cortisol, a steroidal hormone, in the body.[24]

Ex-Wall Street trader turned researcher John Coates showed us how acute exposure to this class of naturally occurring steroid, say during a prolonged period of stress, causes anxiety and selective attention to mostly negative information, people or inputs. This exaggerates threats or risks where they may be small or non-existent. Extended periods of volatility, such as uncertainty in one's career or home life, adversity in relationships, discrimination or even geopolitics can be enough to trigger this spiral and change our decision-making profile. In the same way, collective increases in fear and cortisol levels across boardrooms in response to adversarial market conditions could reduce decision makers' collective appetite for risk, dampen risk taking and investment and so exaggerate the market conditions that caused the anxiety in the first place.

On the other hand, Coates' research showed that on the days when male financial market traders had raised levels of testosterone (another steroid or stress hormone) they produced better financial returns than on other days. Not a surprising phenomenon, given that testosterone increases fearlessness in the face of new situations and hence one's appetite for risk. In financial markets, reward still relies on risk taking and so these hormones serve a necessary role, in moderation of course. However, chronically elevated and persistent levels of volatility tip traders into uncharted territory of risk and sensation seeking, impulsivity, and in some studies[25] has led to irrational risk/reward trade-offs. Testosterone is also implicit in reward and addictive functions. Too much of it, say from a streak of new client acquisitions, a bull market, or even an electoral college victory, leads the recipient to prolong the euphoria through increased, and increasingly irrational risk taking. Typically associated with men, testosterone and its effects are also present in women, but to a lesser intensity.

Without an objective lens through which to view your body's evolutionary and chemical responses to the events that shape your life, it is impossible to disentangle the impact of emotion from your thinking. You cannot eliminate the chemical reaction that occurs in your body, but you can attempt to counter its effect on your risk perception. Some personality types are inherently better at doing this than others.

What I'm really interested in are the more subtle everyday emotional experiences that influence our judgement, such as stress, gut feel and social relationships, which seem to buffet us on a daily basis. In chapter 13 we will explore the idea that we're by no means at the mercy of our emotions, but rather in a symbiotic relationship in which both body and emotions depend on each other for optimal functioning. We don't need to control our emotions or banish them, but a deeper understanding of this symbiosis, and how to nourish and maintain it, will lead to improved judgements—especially under stress and uncertainty, where the role and impact of physiological processes is heightened. We'll explore these ideas in detail later, along with some management guru myth-busting, like the oft-quoted 'amygdala hijack' that is supposed to wreak havoc with our emotional stability.

So, is a good decision one that is not clouded by emotions? Yes and no. Judgements that don't seek to understand and embrace their emotional component don't make for good decision making. Accepting that emotion is the conduit through which we make decisions and looking to understand its influence and the information it brings, leads to improved decision making. More to come on the practicalities of this later.

This brings us to point 4 of our crowdsourced hypothesis. Must a good decision align with our own or our organisation's values? Surely it must?

5

MUST GOOD DECISION MAKING ALIGN WITH OUR VALUES?

Can values be biases that hinder our thinking? Why ethical decision making isn't necessarily about doing the right thing. Try a pop quiz on your values and the role they play in decision making.

Where are we now with validating our crowdsourced hypothesis about what good decision making is?

As a reminder it was that a good decision:

1. achieves its objectives
2. logically considers or weighs all the options at hand
3. avoids thinking clouded by emotions
4. aligns to the organisation's or individual's values
5. avoids regret

We're at number 4. Must a good decision align with personal or organisational values?

Sixteen years ago, I would have answered unequivocally *yes*. My management guru books had taught me that authenticity is a desirable and inspiring leadership trait and the root of authenticity is being consistently true to one's values and hence decision making should be values-led. However, 15 years ago I moved from Africa to Southeast Asia, where I lived for 12 splendid years across Hong Kong and Singapore, working and travelling throughout the region. In the latter, I lectured on several MBA and executive programmes, with participants covering 18 nationalities and age groups spanning two decades. I learned a few things about values in this time. Some were pretty obvious with hindsight: that values are heavily influenced by culture, change over time with the accumulation of life events, can be subsumed by the need to belong to a social group and are easily confused with desires.

I also learned some not so obvious things, such as the fact that most of us can't clearly articulate our values on the spot, values can be context dependent and, importantly, that deep-seated or unchecked values become frames through which we subconsciously filter information. When reinforced for long enough, these frames become beliefs that bias our actions and decisions. When challenged by those who hold alternative beliefs, our desire to defend our values can trump logic as we gather more information to support our view, and so become further invested and entrenched in these beliefs—known as belief polarisation. This seems rather extreme, but from a 2020 perspective, with a Trump presidency in the US, a trade war still raging with China and the full effects of Brexit[26] becoming clearer, it would seem that polarisation of value-fuelled beliefs has a profound impact on a collective level.

Let's examine Brexit—which side do you think made the better decision? Those who voted to keep the UK in the European Union, or Brexiteers who supported the Leave campaign? I'm guessing that you would choose the camp most aligned to your values? In America, would you be a Republican or a Democrat? Conservative, liberal or socialist? Where do you stand on climate change, vaccination, or abortion? Your values will guide your choices on these issues, but because it is right for you, does it mean that you chose wisely? Maybe?

Breaking news in this area is both interesting and troubling; Israeli[27] researchers recently found evidence of why our beliefs are so resistant to change. Acceptance of opinions that confirm our beliefs and agree with our values occurs in a rapid and involuntary manner[28] uncomfortably similar to the processes that occur when we encounter uncontroversial facts. We don't have to think about the facts to make up our minds about them, and we certainly don't pause to question them. A subconscious nod admits unquestioned opinions as facts to the debate at hand. Such a pattern of processing may limit our ability to consider and challenge our previously held views, which is an essential component of rational and constructive discourse and decision making.

Such entrenched values feature heavily in religion, politics and product marketing, but what about values in an organisation? Would good decisions align with these? Should they?

The values-led organisation certainly is a trending topic, as companies trim down their extensive codes of conduct and ethics to more bite-sized and easily digestible guidance. A client organisation recently replaced a detailed dress code that had been in place for over three decades with the guidance that they should 'dress appropriately at all times'. Five weeks later the old dress code was reinstated to the dismay of the workforce, who now resembled a ragtag group of underfunded students rather than the conservative, quintessentially British organisation they had been for the last 150 years. Dress code is hardly a deal breaker, but when an organisation's and individual's values clash then decision making is frustrated.

When it comes to ethical decision making, which is values-led decision making, demanding of employees that they 'do the right thing' and then letting them get on with it is not a feasible solution either. Our values determine what we believe to be 'the right thing to do' and, given the issues highlighted above, unexamined values don't always make for good quality choices. Clarity and enforceability of organisational values with a behavioural[29] definition of what it means to 'do the right thing' is essential for any success in rolling out ethical decision-making programmes.

Ethical decision making

Some cutting-edge work is being done in moving beyond codes of ethics to design behavioural decision-making strategies that aim to improve ethical decision making in organisations.[30] These strategies target individual decision makers by acknowledging that decision making is both emotion and context led. This includes calling out differences in values between employees and organisations and providing a framework that examines these and their impacts across stakeholders (including the decision maker), both immediately and over time. At the core of these strategies sits curated organisational values and explanations of the behaviours that reflect these values in action. Such a behavioural strategy becomes a powerful tool to both engage and empower employees in decision making on behalf of an organisation. Alignment of corporate values with individual decision making seems sensible, provided the values are, too.

Voted America's most innovative company by *Fortune* magazine for six years, Enron was a blisteringly successful commodities, services and energy company employing over 21,000 smart, driven people around the world. Its 64-page code of conduct headlined by making it clear that all employees have a responsibility to conduct 'their business affairs [...] in accordance with all applicable laws and in a moral and honest manner'. At Enron, their values were at the heart of everything they did, and staff were expected to make decisions on the company's behalf within the confines of the following: [31]

Respect: We treat others as we would like to be treated.

Integrity: We work with customers and prospects openly, honestly, and sincerely.

Communication: We have an obligation to communicate.

Excellence: We are satisfied with nothing less than the very best in everything we do.

How could this not work out?

In 2001, Enron failed spectacularly and still holds the record for the largest business bankruptcy ever. Despite a sensible list of corporate values, leadership subverted the law at every opportunity in pursuit of short-term profits and personal gain; this filtered down into the corporate culture and altered the accepted operating values of the organisation. What keeps this story relevant almost two decades later is that it was as much a moral as a financial failure and it has fundamentally changed how businesses are regulated, ushering in a new age of business ethics.

Wayward values and immoral behaviour aside, is it even possible to make decisions that go against your values? Of course, we do it all the time: we succumb to peer pressure and go with the group decision; even if we aren't super comfortable with it, we sometimes sacrifice one value in pursuit of another; or even tell little lies, despite valuing honesty. We may value a world where no child goes to bed hungry or without clean water, yet may still spend our spare cash on nice dinners or flight upgrades rather than making charitable donations. None of this is unusual.

Could you name your values? Maybe just three of them?

If you're struggling, think about your anti-values*—the things that tick you off or frustrate you, such as ill-mannered commuters or a micromanaging boss. What would these anti-values tell you about your values? Perhaps being respectful and seeing the value in others is important to you? It would be uncomfortable, then, to make decisions that bullied or belittled others, regardless of what your employer expected of you. I could never suggest that ignoring your values or going against them in decision making would make for better decisions—it simply wouldn't. Values and beliefs are lenses through which we filter information. They

---

* I first heard of anti-values from a trailblazing career coach and friend, Alison O'Leary.

guide our attention to what we believe is important, provoke emotional responses and the concomitant chemical influence on our thinking when they are breached. They can dictate our risk preferences and profiles at different points in our lives. Ignoring them is a dangerous strategy.

So, is being aligned with our values essential for good decision making? Understanding your values and how they direct attention, provoke responses and influence risk appetite is foundational in choosing wisely. You certainly can't make decisions aligned with your values if you aren't clear on what they are. If you haven't looked them up and examined them in a while, then how do you know they will help you make good decisions under time and resource pressure? You don't.

Organisational values are more clearly stipulated and described. Making decisions that fly in the face of these is never a good career strategy. Understanding why an organisational value exists and how it should feature in an employee's decision making helps bring it to life. Of course, even organisational values should be continuously re-examined in light of whether they still fairly represent the company and its people. This leads us to the last point in our crowdsourced hypothesis: a good decision *avoids regret*.

'There is no softer pillow than a clear conscience' is a French proverb that has stuck with me throughout my life. If we make the best possible decision that we can, with the information that we have under the circumstances we face, then we should not regret our choices, regardless of the outcomes. But how can we be sure we are making the best possible decisions?

PART 2

6

YOUR CURRENT
DECISION-MAKING PROCESS

Your decision-making process is explored and
compared to the components of a best-practice
decision-making framework.

Think of the last important or impactful decision you made.
How did you make it? What were the steps you took? Can you jot
them down?

Most people can—although I'm often asked if this should be a
personal or a professional decision. Many people feel that the
process used in these two settings would be different. It shouldn't
be too different, though. Decisions made on behalf of an
organisation attract a higher degree of scrutiny. You know that
you will have to consult with stakeholders, perhaps show that
you followed a due diligence process and other steps required by
protocol in order to justify your choice to peers and/or superiors.
In anticipation of scrutiny, and to avoid the regret that failing such
scrutiny would cause, we tend to be more thorough in delineating
our steps, gathering our evidence and showing our workings out
in organisational decision making. Decisions in our personal life

tend to attract a smaller group of less demanding stakeholders and so the process used in deciding might not be as transparent or detailed. Where one isn't constrained by a standard operating procedure that dictates decision steps, personal and professional decision processes tend to differ only in transparency and level of granularity.

If you're still not convinced, why don't you compare how you made your last important corporate and personal decisions? Was gut feel more prominent in your personal decision making? Greater scrutiny makes it harder to justify gut feel as an input in your choices, but it shouldn't stop you from understanding where that gut feel came from and using what you discover as a data point in your justification. We'll talk more about how to do that later. Did you take more risk in your corporate decision? Perhaps your risk budget and deployable resources were bigger, and any personal loss limited. This doesn't reflect a different process, but rather a different context. Or perhaps the greatest risk when making corporate decisions comes from getting it wrong in an unforgiving corporate culture and so you choose to take on as little risk as possible, or maybe even toss the decision as far up the corporate ladder as you can? All of these are perfectly normal approaches to making decisions. For the purpose of this exercise, it would be most useful to select decisions that you took ultimate responsibility for.

If you haven't done so yet, please write down the process you used to decide.

Now that you have clarity on your own decision-making process, let's compare it to a best practice decision-making framework and see where you could augment it. This part of the process always elicits a collective groan from my MBA students and pleas to simply hand over the new framework, which they will then duly study and apply on their fast-track ascent to industry domination. For those of you who also prefer to get to the point and skip writing down and reflecting on your own decision-making process, I'll tell you what I tell my students—a tale of lessons learned through years of teaching and facilitating decision making:

'Decision making is personal and as unique to you as your fingerprint. It reflects your values, beliefs and priorities. Your decision-making process is predicated on your unique psychological skill set: accessible mental states, intelligence, memory, confidence, risk preferences, bias profile (including values) and time orientation. It also reflects the resources (physical assets) at your disposal: time, money, access to information, and a social network that supports and advises you. I can give you a social media-worthy *Five Steps to Invincible Decision Making*, but if this doesn't reflect how you engage your physical and psychological assets in actually making a decision, then it will be nothing more than a tick-box exercise imposed upon your own process. An annoying overhead, and the first thing you jettison when under pressure of any kind.'

What follows are researched and tried-and-tested ideas that will improve the quality of your decision making, but only if used to supplement your existing strategy. Pick the ideas that appeal to you and slot them into your own process where you think they would work best to improve decision outcomes.

A best practice decision-making process

A definition of a good decision usually begins with engaging a decision process that allows the decision maker to decide with as little regret as possible. Decision processes are highly personal, but should include some of these success factors:

- a process rather than outcome orientation
- clear decision rights
- a meta-decision (including ensuring that the correct problem is being solved)
- checking that the problem is being correctly framed
- being aware of the mental biases you/your team are most prone to
- exploring assumptions and risk
- understanding the role of unconscious processes on risk perception
- gathering challenging opinions

The definition of a good decision never includes whether the desired outcomes were achieved. There are so many variables that affect the outcome to a decision that a well-made and considered decision can still result in an undesirable or less optimal outcome and vice versa. Let's explore each of these in turn, allowing you to decide which ones will add the most value to your current process.

7

PROCESS VS GOAL ORIENTATION IN DECISION MAKING

Praise for process orientation in decision making
and what not to do if your son takes your Porsche
out for a spin without asking.

A 2010 McKinsey[1] study analysed 1,048 strategic decisions made by their clients in areas ranging from mergers and acquisitions to organisational change. Their success criterion was return on investment (ROI) and, using regression analysis, they explored which elements of decision making contributed most to it. Beginning with the area that companies (and individuals) initially focus on when making strategic decisions—gathering good quality data—and subjecting it to the best possible analysis to produce predictive and scalable financial models.

Their results were somewhat surprising. They found that *data quality and quantity* only contributed to an 8 per cent increase in ROI. Idiosyncratic variables such as capital availability, investment opportunities, and market sentiment, contributed 39 per cent, but the bulk of the impact came from the quality of the *process* used to exploit their analysis and reach a decision. Examples they provide include: explicitly exploring major uncertainties; ensuring participation in discussions by skill and experience rather than rank; and soliciting and including perspectives that contradicted senior leaders' points of view. Raising the quality of a company's decision-making process from bottom to top quartile on these measures improved ROI to a company's investments by 6.9 per cent. Not a trivial contribution at all.

Building a decision process is something most of us, and most companies, haven't really considered. Instead, each decision is explored on its own merit according to the desired outcome or objectives. We also tend to judge decisions by their outcomes because outcomes are easier to measure. Of course, poor outcomes weigh more heavily in any evaluation than good outcomes. In addition, we're usually only rewarded for those good outcomes because the latter is assumed to result from a good decision process. In reality, this *outcome focus* skews risk perceptions and results in weaker decision making over time.

Our time together to work on improving your decision outcomes begins with a shift from goal to process orientation. However, if you've only ever judged a decision by its outcome, then moving to a process orientation may not be an easy mindset shift. We'll take it slowly, starting with some celebrity inspiration.

Process poster boys

Most decisions involve some uncertainty. This places them in the realm of bets and gambles, terms more frequently associated with games of chance and investments. In this context, evaluating a bet as good or bad would depend on the stakes and the odds,[2] not the outcome. It is no surprise, then, that successful investors focus on building strong and sustainable investment processes. In fact, two of the world's most successful investors, Warren Buffett (Berkshire Hathaway) and Ray Dalio (Bridgewater Associates), are poster boys for process-driven decision making. An investment process captures the decision criteria and processes that guide an investor when choosing to buy or sell assets for their portfolios. Buffett's process lives on in Berkshire Hathaway and lays down the steps required to delineate a good company from a good investment.

His process limits his analysts to investing only in companies they understand deeply and can analyse thoroughly. This might seem obvious, but is something that is increasingly hard to do, as companies spawn across geographies in various corporate structures, management turnover is increasing rapidly, and strategies are becoming ever more short term. This process resists the fashion of trading shares on stock price or news flow to make a quick buck. It also dictates that they initially analyse a company independently of the market, sentiment, politics and price. 'Price is a frame that affects one's evaluation of a company,' says Buffett.[3]

Understanding the level of rationality in a company is not a feature you would find in many investment processes. In this case, rationality explores whether[4] management is selfish with cashflow and profits or uses it to the benefit of shareholders. Do they follow competitors' strategies blindly or have the ability to unlock value through unique positioning or offerings in the long term? A total of 12 tenets are said to underlie his process, including numerous financial metrics and ratios. I should add that they seem overly neatly packaged in the media—perhaps for public consumption—whereas the reality may be somewhat more fluid and less structured.

Ray Dalio built one of the world's most successful hedge funds at Bridgewater Associates. His investment process has a controversial behavioural aspect to it that he calls 'radical transparency'. It requires every employee to be completely transparent in their thinking and opinions and gives everyone in the organisation access to recordings of closed-door investment conversations (within regulatory and legal limits). But if everyone has an opinion, how do you know whose is more important in your decision making—or do you just block it all out and go with your own? The solution, according to Dalio, is weighting opinions according to someone's believability in the area under discussion—and he's even had software developed to do this. For example, if a financial stock is being discussed, everyone is entitled to an opinion; but the financial analysts, and specifically the most successful among them, will have their opinion upweighted. There are mixed reports on the mental toll that working in such an environment takes, but the investment returns have been spectacular.

Transparency and challenge are two components that have featured in all decision-making processes I've explored with teams. The degrees to which these feature and the shape they take depends on the goal of the process, be it driving consensus or generating robust debate that allows individual decision makers to fully explore an issue before deciding. Whatever the goal, the first step is to construct a process to house these protocols. Such a process is more likely to produce good outcomes over the long term than a results-orientated approach to decision making—especially when specific decision objectives are unclear or changeable, or external influences are unpredictable.

Outcomes skew risk perception and evaluation

What is a results-orientated approach? While it is clear that process orientation emphasises the development and constant tweaking of a decision process used to explore problem domains, it must still achieve an objective. Within such a process, the objective is to make the most robust decision. Here, solutions are shaped by the exercise of understanding the problem domain. This allows for goals or objectives to change if this exploration of the perceived problem uncovers an incorrect diagnosis or new factors come into play.

With a goal-orientated approach the goal is fixed, and the process focuses on achieving it. Such processes can be random, constantly change or follow the age-old 'What do we need to do?' and 'How do we do that?' team discussions. Goal orientation is action orientated, solution driven and less reflective and exploratory, but quicker. I'm not suggesting you announce to customers that your staff are no longer goal driven! That would go down like a lead balloon, because what customers wouldn't know is that process orientation in decision making allows for a better understanding of the challenge you are addressing and greater agility in the face of rapid change.

Sticking with our finance examples, one of the issues plaguing the investment industry is compensation levels. The bursting of the tech bubble in 2000, the 2007 subprime crisis and subsequent financial crisis from 2008 onwards all occurred at a time before stringent regulations on employee's compensation was implemented, when the goal for each investment year was to maximise one's bonus. Fair enough, given how short investment careers were and the accompanying stratospheric stress levels. Taking short-term bets with increased volatility or trading on news flows or sentiment is a quick way to ramp up short-term returns, especially near financial year end. Lurching from bonus to bonus doesn't help build long-term stability in a portfolio or investment business. Incentivising investors on the quality of their risk-weighted returns over a three-to-five-year period will change the nature of the risks they take, removing some of the stress and accompanying knock-on effects of a bad period, while encouraging longer-term relationships with the companies they invest in. In the same way, incentivising a sales team on new business generated produces an outcome focus rather than creating incentives to invest in client retention and relationships.

Building and debating decision processes is commonplace in investment management, but outside of this process-driven industry discussing the process underlying one's decisions is not typical canteen conversation. Usually, when our own investment portfolio falters, we turn to our financial advisor or mutual fund manager to explain to us what went wrong. We expect him or her to highlight how market sentiment, global upheaval or other exogenous events have impacted our returns. We seldom prod our advisor's decision-making process and how that may have resulted in our portfolio pickle in the first place.

Researchers concur that we struggle to judge a decision by the process used. As soon as the results of a decision are known, this becomes the focus of any evaluation. A surgeon cannot claim an operation was successful if the patient died on the operating table, despite faultless decision making and skill. This devastating outcome overwhelms any other data about the surgeon's choices. In all decisions, consequences are far more noticeable than the process that produced them. How much weight is given to consequences can depend on who is doing the judging. Who is judging the surgeon's choices in the operating theatre? A surgical colleague might consider the surgeon to have performed admirably, but the deceased patient's loved ones might never accept a diagnosis of success. Outcomes also occur closer in time to the act of judging a decision and so have more influence on our judgement than the process that produced it—this is known as the 'fallacy of saliency'.

It's also been shown that knowledge of an outcome can change perceptions of a decision made before the outcome was known. We've all done this. Rami Malek won a 2019 Oscar for his portrayal of Freddy Mercury in the semi-biopic blockbuster, *Bohemian Rhapsody*. In his acceptance speech, he mentioned that he hadn't been the first choice of actor for the part, but with his Oscar in hand he smiled and added, 'but I guess it worked out OK!' I'm also guessing that the producers of the film were congratulating themselves on their excellent choice, even if they had been doubtful before.

In a typical work environment today, it's usually only the outcomes of employees' decisions that are observed and judged. If the 'decision judge' and 'decision maker' use different frameworks to evaluate decisions, then the one that pays the bills is going to be more influential. Decision processes take time to develop and use. If organisations don't consider these processes in judging and rewarding decisions, then they will continue to suppress decision quality. As an advocate for decision processes, I am often faced with the question of how to judge a decision. Should managers or HR professionals be experts in decision making? Is it even possible to judge a decision's quality?

Judging your own choices

Actually, being a good judge of someone's choices is one of the oldest professions. Who would do such a job? Judges, of course. Every crime begins with a choice (either premeditated or not) to commit to a certain course of action. Imagine if a judge delivered a verdict after only hearing the details of the actual crime committed (i.e., the decision outcome). For example: Mrs White killed the postman with a blunt kitchen utensil. Murder is against the law therefore Mrs White must be guilty.

Despite reducing court cases down to minutes, such simplified judgements would be unacceptable. We expect judges to consider the context in which an act was committed, the circumstances that led to the act, including how much information was available to the accused, the emotional state of the perpetrator and how that impacted decision making. In the eyes of the law, in most judicial systems, murder is wrong, but how and why a murder is committed will determine the appropriate punishment. Was it premeditated or committed in self-defence with a clear mind or under emotional or mental strain? Fortunately, case studies and the law serve as criteria against which to judge these complex choices. In our personal decision making, values are often called upon instead. So these tend to become meaningful criteria against which to judge and justify our own choices. Not every decision has a moral aspect, though, and when we lack such 'credible' criteria against which to make a judgement, the *outcome* becomes the gold standard.

Bearing in mind that when we judge decisions against their outcomes, we do so under the influence of many mental biases, one regular rascal being loss aversion. Even if we are judging someone else's decision, negative outcomes will weigh more heavily in our thinking than a positive outcome.

If your teenage son nicks the car keys and takes your prized Porsche out for a midnight spin with his friends, two of many possibilities could result:

1. he returns it without incident, but you discover his betrayal; or
2. he has an accident that damages the car.

Now imagine that you discovered he had taken the car before he returned. You would probably go ahead and make a judgement about his decision, foster an appropriate level of ire and, possibly, devise a punishment. All before he returns. After he arrives home and shows you the damage inflicted from the car accident, would you apply a different judgement? Most likely. In which case are you going to be angrier: taking the car without permission or taking the car without permission and reversing it into a construction barrier? If the crime is taking the car without permission, then the added injustice of an accident shouldn't change the punishment. But it will, of course. Decisions that produce poor outcomes (a loss rather than a gain) are far more impactful and newsworthy and so tend to be judged more harshly.

What does this have to do with anyone adopting a decision process? Well, you're the ultimate judge of your own decisions, you have to live with the consequences. If you can reflect on and judge your decisions in terms of the process you used to generate them, you will find it much easier to move away from focusing solely on the outcome. An outcome focus encourages the decision maker to fixate on a desired goal and work back from there to gather information and explore alternatives in less structured ways. This fosters mental biases, blind spots, lopsided risk assessments and a difficulty in standing up to scrutiny.

In summary: why do I propose developing and using a process in decision making? To uncover our blind spots, allow for our limited memory and processing capacity, rightsize the impact of emotions and counteract thinking mistakes. Ultimately, the use of an appropriate decision process should render the act of deciding trivial. Sounds sensible, doesn't it?

Let's turn to the first principle recommended for a robust decision process: allocating decision rights.

8

DECISION RIGHTS

Who is the decision maker and why does it matter?
Engineering fluid decision making in the police service.
What are decision rights and risk budgets?

'I know it's not the best time for them to be doing this, but they'll stay south of the River Thames like they promised,' he replied to Detective Chief Superintendent (DCS) Kevin O'Leary. 'We are the world's oldest democracy and our job is to maintain order, not decide who can express their right to be heard. Besides, they've been doing this once a month for 18 years. We know the drill by now.'

He had a point, but it was the day of the opening ceremony of the 2012 Summer Olympics in London. DCS O'Leary (his real name) was responsible for all detective, intelligence and forensic operational teams for the London Olympics and didn't fancy diverting resources to face down hundreds of cyclists who had decided to exercise their right to make a statement, today of all days. What the cyclists probably didn't appreciate was that 40,000 security personal were mobilised across the city in the largest peacetime security operation the UK had ever mounted, deploying more armed personnel than those stationed in Afghanistan at the time. Ok, so they had to stay south of the River Thames and not cause too much obstruction to traffic. What could go wrong?

It was a warm evening and London was abuzz in the hours before the opening ceremony. Surveillance cameras tracked the cyclists as they meandered towards the Thames, their numbers swelling as they rode. All at once it became clear that some of them had broken formation and were heading onto a bridge along a path towards the Olympic Stadium. Others soon followed, mounting the bridge at exactly the moment the Queen's cavalcade passed underneath. This wasn't the plan, and images of the previous year's riots that saw Prince Charles and his wife separated from their security detail and mobbed by angry rioters, surfaced effortlessly in O'Leary's mind. Tonight, 342 million people were watching London as she sparkled and heaved with tourists and spectators, ripe for a major incident to unfold—and this could be the trigger.

There were only a few options available to contain the cyclists and almost no time to implement any of them. They could form a barrier of police officers to physically stop them from entering the Olympic zone, but that would take too long and most likely only redirect rather than stop them. They could arrest them—but there were now almost 200 cyclists on a peaceful mass gathering, which wasn't strictly illegal, just very, very inconvenient. Someone had to make a decision in the moment and accept full responsibility for the consequences. O'Leary knew that he had that authority and so didn't hesitate to act. 'Arrest them. All of them,' he ordered.

That would have made a perfect Hollywood ending, but the reality was a bit different. O'Leary had the authority to take that decision, but along with it came a duty to ensure that he understood and accepted the associated risks and that others had the resources required to implement his decision and contain any potential fallout. His decision making didn't end with the choice to arrest the cyclists, nor did his accountability. He would be reviewed by peers, authorities and in a court of public opinion for this decision, his decision-making process and ability to empower others to implement his choice.

Arresting 182 people on a busy night was no mean logistical feat. Enough holding cells had to be located to contain them, enough police stations in the area with capacity to process them and enough buses to transport them all there. But it was done in short order because everyone under O'Leary's command understood the role they played in the operation and didn't need to seek permission to exercise their decision rights.

In emergency situations, there is no time to decide who should decide; yet it is absolutely critical that all emergency responders work towards a common objective, knowing exactly which decisions they are at liberty to take and which should be referred on and to whom. Who takes what decisions is usually well-defined within any government body, but when a major incident breaks out, such as a riot, the person of the right rank isn't always in the right place with the right information at the right time to make the necessary decisions. This can paralyse emergency response services at worse and delay them at best, which is what happened when a riot broke out in North London in October 1985. Confusion around who was able to take decisions led to a recognition that tactical roles are more important than rank in an emergency and that decision and command structures must be fluid to be effective.

In order to achieve such a fluid system of command, the UK's Metropolitan Police Service developed the gold–silver–bronze command structure. A bronze commander is directly responsible for all operational decisions and control of resources at the scene. The silver commander is the tactical commander who, following the strategic direction given by gold, is able to set priorities and objectives for bronze command. Silver may or may not be present at the scene. The gold commander is never on site but rather stationed at a distant control room unsurprisingly coined Gold Command. Here, he or she will formulate the overall strategy for dealing with the incident. What makes this system of leadership especially effective is how these command colours are allocated. The first blue light responder of any rank at the scene takes charge initially and their radio becomes the one used for communication, while their car's blue lights are left on, indicating its use as the forward control point. As soon as a more senior officer arrives, the bronze command is passed on, along with a vest identifying the new tactical commander. However, the first responder who assumed initial control stays with the bronze command to ensure continuity.

You'll know you aren't in the police service because tactical decision making in your organisation is probably not as well defined or executed. Are decision rights clearly delineated? Are risk budgets allocated to match those decision rights? Does everyone take the necessary decisions in a timely way, working towards a well-understood, common goal? Are heat-of-the-moment decisions debriefed and scrutinised to extract learnings and improve decision agility?

By the time I am called in to work on improving an organisation's decision-making capabilities, an engagement survey or something similar has highlighted several issues with their collective decision making. These challenges can include decision making that is:

- slow
- circular
- out of touch with market needs
- insular or siloed

It often involves decision makers who:

- are removed from market needs
- don't listen to the experts within their organisation
- don't trust others to make decisions
- micromanage or sabotage the decision-making efforts of their staff
- don't effectively communicate the context and rationale underlying decisions, leaving others to guess

Do any of these wrongs sound familiar? You could probably list a few challenges that you've picked up when working with others or, dare I suggest it, may recognise some of these in your own decision making. There is only one way to improve decision making in any organisation and that is not to gather more or better data or build better predictive software, but rather to improve the decision-making smarts of the people who use the data or software, therefore improving outcomes.

Once you know you are solving the correct problem, it is important to be clear on who the decision maker is and ensure that they have both the authority and resources to exercise their right to take decisions. Like operational command at bronze level, each decision maker should know how much risk they can afford to take with the budget and other resources allocated to their decision-making efforts.

Once we know who is deciding, the next step is to decide how to decide and for that we have a meta-decision.

9

THE META-DECISION

Deciding how to decide. Choosing between
winning a battle or ending the war.
A numbers problem in sheep's clothing and
what drug producers are learning from Netflix.

"Nothing is more difficult, and therefore
more precious, than to be able to decide."
—Napoleon Bonaparte

Do you know what a meta-decision* is?

It is the simple act of deciding how you will decide *before* you jump
in and gather information, make a decision or solve a problem. It
begins by checking that you are, in fact, solving the *right* problem
within the right frame; then asks you to decide *how* you will solve the
problem—with what tools, timeframe, information and resources
and against what criteria. It sounds like a mini project plan because
it is. The meta-decision forms the foundation of a good decision
process—it helps you to anticipate challenges, use the best tools,
and gets all your team members (if any) on the same page. All this
speeds up making and implementing your decision.

---

* I first learned about meta-decisions from reading *Winning Decisions* by Edward Russo
and Paul Schoemaker. An old book but a goodie!

Einstein is widely quoted as having said: 'If I only had one hour to save the world, I would spend 55 minutes defining the problem and five minutes finding the solution.' As much as I would have liked the great man himself to have said this, it appears to be a misquote based on a collection of articles published in 1966 that included a comment by the then head of the Engineering Department at Yale:[5] 'If I had only one hour to solve a problem, I would spend up to two-thirds of that hour in attempting to define what the problem is.' I imagine the original is so widely misquoted because it sounds smarter when it comes from Einstein and because no one really bothers checking quotes out—especially when it's pasted across an iconic picture of the smiling wild-haired professor. It's also so widely quoted because every good decision maker or problem solver knows how important it is to spend more time thinking about the problem than the solution.

The steps included in a meta-decision are:

- solving the correct problem
- framing
- methodology, resources and time allocation
- similar experiences

As a consultant to consultancies, I'm continually embedded in different teams across different organisations. One team in particular would begin each weekly call with a 'status update', 'objectives for the week' followed by 'next steps'. Tasks would then be doled out and everyone would head off and get them done. Such an action-orientated team is great to work in, especially when dealing with a linear objective, like organising an event or developing a project plan. This particular team was designing a decision-making strategy for 7,000 employees scattered around the world in a highly regulated industry. It was neither a linear nor action-orientated project. It required sharing of expertise and experiences, strategy, research, debate and reflection, alongside their day-to-day duties. But who has time for all that?

Continuously preferring action over contemplation resulted in us going around in circles, wasting resources and getting pushback from the business. We weren't clear on the overall impact we wanted to make and the system within which this project took place. Each team member was very focused on exactly what they had to do, what they had control over and what they could tick off their to-do list. To be fair, they were extremely busy, and this project had been thrust upon them. I knew it wasn't going to get better until they could answer one simple question: What is the problem you are trying to solve? Despite already being knee-deep in solutions, all eight members of this multinational working team gave me a different answer. Escalating this to the project sponsor a few levels up helped us understand the corporate rational for the project. From that we were able to understand the problem we were supposed to be solving, how it had been tackled in the past, what resources were available, what were the expectations from senior management for when this should be done, as well as devise acceptable objectives.

With deadlines and timelines all creeping in on us, it's not unusual for teams to be highly action orientated. After all, doing stuff gets stuff done! This is fine if we are sure we're doing the right thing, but as decisions become more strategic and impactful, there's no getting away from investing time in thoroughly exploring the problem domain. A handy tip here is to separate decision meetings from action meetings, in order to reduce our innate bias for taking action. This allows you to protect time for thinking and create separate time for doing or allocating tasks—even if a one-hour meeting is divided into two half-hours by a coffee break, a clear separation is all that's needed to avoid one activity bleeding into the other.

This is where a meta-decision can help bring structure to decision-making conversations. It consists of five steps in no particular order, although I would recommend starting with step 1.

Meta-decision step 1: What is the actual problem we are solving or objective we want to achieve?

In World War II, the Axis Powers (Germany, Italy and Japan) had a sophisticated communication system in which Morse code was enciphered using the Enigma ciphering system. Cooperation between French, Polish and later English cryptographers at Bletchley Park enabled the Allies to read substantial amounts of coded enemy radio communications enciphered with these Enigma machines. The efforts of these codebreakers, including Alan Turing, resulted in the first programmable electronic computer—aptly named *Colossus*. The intelligence generated by *Colossus* from German radio and teleprinter transmissions was considered more vital in the war effort than any other intelligence and so couldn't share the top-secret classification but was instead classified as ultra secret. In fact, it was so secret that the existence of Bletchley Park and its codebreakers was only declassified in 1974.

Once German naval messages were successfully decoded, the Allies had information on where and how German U-boats would attack Allied forces. The biggest challenge was not to scramble

Allied forces and prevent these attacks from happening, but rather to let some of them happen. Ensuring that German forces never discovered that the Allies could intercept their communications was a top priority. Using this information strategically was the goal—not to defeat the Germans in individual battles but to allow the Allies to engage in strategic manoeuvres that would eventually win the war. Sometimes they would instruct Allied ships to move away from a planned German invasion and other times they let the battle rage. Not every codebreaker was comfortable with knowing that some attacks were allowed to happen. Living with this knowledge was a feat of superhuman restraint, made bearable by knowing that if they succeeded lives would not have been sacrificed in vain. They knew that ending the war—not winning individual battles—was the objective. Their intelligence was said to have shortened the war by at least two years, saving countless lives.

Hang on, what if you aren't at war? What if there isn't a problem and things are going so well for your organisation that you simply want to capitalise on that success? Defining the core objective of a decision or new challenge works the same as defining the problem.

Meta-decision step 2: How are we framing the problem or core objective?

Socrates was credited with being the first to propose that all information occurs within points of view and frames of reference and that all reasoning proceeds from some goal or objective. In 399 BC, the great man was executed for corrupting the youth of Athens by suggesting that authorities were not infallible in their thinking. They, too, should be questioned to uncover the motives behind their rulings. Today, we know and understand the fundamental truth in this reasoning and how it separates good decision makers from the rest. Without fail, every piece of information that is presented to you is done so through someone else's frame, filtered through their own intentions or desires. In the same way, *you* frame all information you offer to others. What is a frame?

Below is a list of numbers from one to ten. There is a sequence in these numbers. Those of you who are good with numbers will have no problem identifying this sequence. If you're not so good with numbers, give it a try anyway.

8 – 5 – 4 – 9 – 1 – 7 – 6 – 10 – 3 – 2

Got it yet? Most people find this sequence tricky or just plain impossible. I'll give you a clue: it's a common sequence that most of us work with every day.

Does that help? If you are ready for the answer, read on …

This problem is presented numerically, which makes you think in numbers. This is reinforced by my mentioning that those good with numbers should find it a breeze—very sneaky of me! In fact, these numbers are in alphabetical order, which is hard to see if you are in a numeric frame of mind. Of course, you see it now!

How a problem is presented, or framed, can influence how we process the information as much as, or more than, the facts of the matter. If we explore a problem through a narrow frame or only one or two frames when there are several more at play, then we are not going to make the best decision we can.

While business dealings in the UK are required by law to be legal, decent, honest and truthful, no such constraints apply to politics or politicians. Stoking fears of lazy immigrants pouring into the UK from Europe and creating all sorts of mischief was a major campaign tactic by the Brexiteers[6]—dubbed 'Project Hate' by some. Taking back control of the border and reducing immigration numbers to acceptable levels was the promise (like Donald Trump's border wall argument but without the drugs and bad hombres). Before the Brexit bell even rang, immigration numbers from the European Union (EU) had fallen dramatically. The latest figures[7] from the UK's Home Office (Feb 2019) tell us that EU net migration has fallen to a level last seen in 2009. Jackpot! Clearly this was a successful campaign, producing results even before Britain regained official control of its border. Brits will no longer have to worry about troublesome foreigners stealing their jobs—that is, if you are comfortable with how narrowly this problem was framed.

The UK has always claimed strict control over non-EU immigration and while EU members are returning home, non-EU net migration was the highest since 2004. In total, immigration is up and will likely continue so even after Brexit. I know it's a bit naughty of me to make an example of politicians' poor decision making, but the success of this campaign shows how readily this narrow frame was adopted by the voting populating when a simple Google search on UK immigration trends would have revealed the fuller picture—that EU immigration is just one part of the picture and certainly not the main driver of immigration numbers. Perhaps they were solving the wrong problem? Is the problem spiralling immigration, lack of retraining as the UK's traditionally labour-intensive industries modernise, or ignorance?

Uncovering frames and hidden agendas can be done by asking questions such as: How is the problem framed? From whose perspective are we exploring it or the information gathered? How wide/narrow is our frame? What are the goals and objectives that I bring as an individual or team to the problem? When you explore information, stay curious about who prepared it and what their frame or objective might have been in doing so.

Meta-decision step 3: How are we going to solve this problem—methodology, resources and time allocation?

You've seen both adults and children alike jump right in to solve a pressing task without first thinking about how they are going to go about solving it, or even if they should solve it. When I ask groups of experienced decision makers to build a device that can safely land a raw egg on the conference room floor from a height of two metres, 90 per cent of them assume they have to build some kind of parachute to do so—an egg shuttle.

Surprisingly, most groups simply jump in and experiment with the materials I've supplied. Planning or prototyping is usually not considered nor is looking for ways in which this has been done before without needing a cleaning crew in the conference room. Almost no one asks if they have to use only the materials supplied and no one has ever attempted to solve a different problem, such as making the landing surface soft enough to support a falling egg. This is an easier problem to solve—by turning a standard conference room chair upside down and making a 'trampoline' with the balloons supplied (I supply a goody bag of materials), or stretching a sweater across the four legs of the chair and voilà, your egg will be caught and supported if enough flex is allowed. In six years of conducting this experiment with participants around the world, this particular solution has never been presented, despite being the first YouTube video that pops up if you google 'how to drop an egg without breaking it'. But that's not what surprises me most but rather that executives, who are by-and-large professional decision makers, don't have a framework for thinking about how they will make a decision or a strategy to guide them—they simply jump in and get busy with finding a solution.

Of course, landing an egg from a height of two metres without splattering it is hardly the type of challenge most leaders face back at the office, so they can be forgiven for taking this less than seriously. However, this exercise is designed specifically to test group decision-making dynamics and processes. Once the group has decided what problem they are solving, they should look to understand the skills and talents in their team and assign duties accordingly. Understanding the materials available, allocating tasks and designing a prototype should all follow, in theory; but picking up the balloons and pipe cleaners and just giving it a go is definitely the preferred method but has yet to yield success in this exercise. Most teams also don't finish in time because time wasn't factored into their plan.

Meta-decision step 4: What lessons can we learn from others?
I know, your problems are unique. No one else is at the cutting edge of x technologies and so there's no point in learning from history or other people's mistakes—you just have to forge ahead and make new mistakes. Not so fast. Could it hurt to prod around and see where others or other industries have faced similar problems?

The market for antibiotics of last resort is in a pickle. When a bacterial infection fails to respond to commonly administered antibiotics, doctors resort to a backup plan of novel antibiotics as a last line of defence. To ensure that bacteria doesn't develop resistance to these, too, they can only be used sparingly, when all other antibiotics have failed, and for a short period of time. These drugs are therefore not produced in large quantities, while prices remain low in line with government pricing policies on antibiotics. (Patients are used to paying up for cancer treatments, but antibiotic prices are kept low in many countries). Low consumption coupled with little pricing power conspire to keep most large pharmaceutical companies well clear of producing these antibiotics of last resort and investors stay away from start-ups willing to take their chances on them. Yet the world desperately needs these drugs, which can cost up to US$2 billion to bring to market. Looking for solutions to this pickle has turned up the usual suspects of offering larger grants and increasing the prices of these drugs (given the current debates around drug pricing this isn't wildly popular) and another solution borrowed from an industry with very little to do with science: Netflix.[8]

Netflix has 139 million subscribers around the globe each paying a monthly subscription fee. So, if subscribers binge-watch their favourite series over a weekend and then watch nothing, or head over to Amazon Prime's video streaming service for the rest of the month, Netflix doesn't see a dip in its funding. It maintains a steady stream of funding for new content. What if healthcare providers paid a subscription for access to novel antibiotics, regardless of the volume they used? This ensures that, when the antibiotic is new and volumes low, the pharmaceutical company still gets paid. Once the drug is more widely used, the price doesn't go up. Such subscriptions are already being tried in America to pay for hepatitis C drugs. While this won't solve drug resistant infections, it would provide an incentive for drug companies to manufacture drugs that doctors would only use sparingly.

Maybe another team in your organisation has overcome similar challenges to you, or aspects of the challenges you face? Maybe another industry, organisation, country or community has solved a shared problem in novel ways. If they haven't quite solved it, maybe their failure holds valuable lessons that could save you and your stakeholders time and money? It can't hurt to take a look.

In their book, *Winning Decisions*,[9] authors Russo and Shoemaker chide amateur decision makers for spending most of their problem-solving time (75 per cent according to their research) on gathering information and coming to conclusions at the expense of understanding frames, thinking about how they will decide and learning from experience—both their own and others. Noting that a carefully constructed meta-decision can save time and money.

If you have colleagues or family members (especially teenagers) that bolt after every good idea or jump straight in to solve problems without pausing to think of the best way to do it or even if it should be done at all, it would help them tremendously to share the idea of *deciding how to decide* first in a few easy steps. If, instead, you know someone who can never make up their mind or is afraid to commit to a certain course of action, the meta-decision process will bring much needed security and structure to their thinking as well.

This step is really about checking your thinking and rooting out assumptions. Once you are clear on the challenge or objective, then seek clarity on the tools at your disposal, information and resources available, who should be consulted and over what timeframe. I've seen many teams leave an action-orientated meeting knowing exactly what it is they have to do but having to circle back later because they found that they haven't consulted, consented or informed the right people or don't have sufficient time, decision rights or risk budget to tackle the task.

What are we assuming about the resources or time we have and who needs to be consulted? This is a good question to ask, along with: Where has a similar problem been tackled and what can we learn from it?

10

THE BUSINESS OF BIAS

The basics of bias. The reality of being biased.
The futility of bias training and the names of
biases that will make you sound smarter.

Let's recap where we are in our decision-making best practice framework. Remember that decision processes are highly personal but should include some of these success factors:

- a process rather than outcome orientation
- clear decision rights
- a meta-decision (including ensuring that the correct problem is being solved)
- checking that the problem is being correctly framed
- being aware of the mental biases you/your team are most prone to ← *we are here*
- exploring assumptions and risk
- understanding the role of unconscious processes on risk perception
- gathering challenging opinions

We are at the very heart of the process now, where we must turn to the engine of all our thoughts and actions—our brains.

Our bountiful brains are extremely resource intensive and can consume up to 20 per cent of the body's glucose supply and oxygen. Fully aware of its own appetite, our processor will try to conserve energy whenever possible, using an array of energy-saving devices. Mental biases and heuristics, or shortcuts, are such devices. Not only is our brain a fuel guzzler but it's also a slower processor than we'd like it to be. When I ask audiences which they think is a faster data processor—their conscious brain or a deca-core computer (one with 10 processing cores)—I am still amazed that the majority pick the conscious brain as the faster of the two. Would you have picked the brain? We'd like to think that we walk around with something that can't be replicated by a machine and definitely not improved upon. But if I gave you a spreadsheet with a hundred figures to be multiplied together, would you say, 'Oh, don't bother with Excel, I can calculate it faster in my head.' No? I didn't think so.

We may be able to perceive and process a variety of more *subtle* data than computers, but when it comes to raw processing power our brains chug along at a fraction of the speed of an old 56K modem. Remember those? The only thing that I remember about them was having to dial up to some far away server and then wait for a tiny trickle of bandwidth to connect me to the slow-motion magic of the World Wide Web. Our brains process information even slower, way slower. Part of this bottleneck sits in our working memory. Working memory is the unsung hero of intelligence, the part of our short-term memory that we call on for immediate perception, information and language processing. It can be compared to your computer's high-speed memory that stores data or bits of the programs that you are currently working on so that it can access them more quickly. The size and processing speed of our working memory is closely tied to our ability to reason and solve problems, yet scientists and users alike report two rather fundamental flaws in our working memory: it has very limited capacity and leaks like a colander. You may have heard that we can only hold five to nine pieces of information in working memory. I used to dismiss this as an urban legend and insult to human intelligence, but I was wrong on all fronts.

It would seem that adults can indeed only hold a limited number of independent chunks of information in our working memory. Neither five nor nine is a big number but the difference between being able to work with five or with nine bits of data is like shopping in the UK property market with £500,000 or £900,000. The difference in what you can buy with these two amounts is vast—£500k would get you a two-bed flat in London, whereas £900k might get you a four-bed semi-detached house. A larger working memory capacity means that more variables can be considered at the same time. Given its size limitation, information stored in working memory must leak or decay to make way for new data. Ever walk to the fridge, get distracted en route and then forget what you came for as you peer at the starkly lit milk and margarine? That's working memory decay in action.

Eyewitness testimony uses long-term memory and the ability to reconstruct fragments of information gathered under heightened stress. Information that would have had to pass through working memory on the way to becoming long-term memory. For some time now, researchers have questioned the reliance on eyewitness testimony in criminal convictions. Their scepticism is duly supported with the results of DNA testing, which has exposed our inability to accurately record and recall information when we're not expecting to have to do so. In fact, a staggering 73 per cent of the 239 criminal convictions overturned before 2008 in the United States through DNA testing* were based on eyewitness testimony.[10] That's 174 innocent lives ruined by someone's perfectly normal memory.

Identifying the bad guy might not be on our agenda, but relying on information stored in memory to make decisions usually is. Yet some people appear to make good judgements in complex and information rich environments really quickly, such as the paramedic who must weigh up a tremendous amount of information and make a life or death decision rather quickly. Same for a fighter pilot, soldier or firefighter. Or the CEO that is under pressure at a board meeting to deliver a quick insight on a complex topic. Rather than a larger mental motherboard that houses more short-term memory, each of these professionals have a repository of past knowledge, experiences and decision outcomes feeding into their decision-making process at speeds much faster than our prefrontal cortex (the conscious information processor of the brain) can process. Such information can be bundled under the label of gut feel, intuition or, as some researchers have labelled it, 'somatic markers'.[11] In the case of a soldier or fighter pilot, they have also honed and trained their decision making through many hours in a simulator or combat training to reduce reliance on working memory and conscious information processing. In much the same way, seasoned executives have a wide range of painful and successful experiences feeding into their reasoning process and, hopefully, some introspection into the thinking that produced those experiences as well.

---

* DNA testing was first introduced in the 1990s in the US.

Unfortunately, good *gut instinct* along with a large repository of decision experiences and training aren't enough to overcome the limitations of a slow, resource-intensive and working-capacity constrained brain. To help us with this, we have developed mental shortcuts (or heuristics) as part of our thinking process as well. A *stereotype* is such a shortcut. We are able to gather enough information to make a decision about whether we will trust someone, or not, through a single glance. Someone's clothes, hair, tattoos, cleanliness, facial expression, accent and any other markers that we are able to perceive generate a remarkably comprehensive profile of someone in our own minds. It is a snapshot that may represent some truth or none at all. It doesn't take very long either, as conscious inputs combine with subconscious data (beliefs and experiences) in our cognitive workspace to form a thought that nudges us to act on someone's appearance without ever having spoken a word to them.

I once spent a week trying not to stereotype. This involved not forming an opinion or a 'feeling' about someone when I met them for the first time. Do you think I could turn this heuristic off and resist judging at first glance? No, I couldn't. Despite my best efforts, I received internal intel on every new person that I met, including those who were just in my line of sight on my morning train commute, before I'd had a chance to make a considered judgement about them. The best I could do was override my initial flash assessment of someone using much slower rational thought later on. This was rather exhausting though, so I gave up rationalising my opinion of strangers pretty quickly and reserved this intensive process only for people I met in meetings. The young chap on the train with a misspelt tattoo and belt failing to keep his trousers over his underwear would forever be someone for me to avoid.

Behavioural economics explores the mental processes we rely on to make everyday decisions. Supported by advances in neuroscience and neural imaging, it focuses on the mental biases and shortcuts (such as stereotyping) that we generate as we process data. There are currently over 220 researched, peer reviewed and documented biases in the behavioural economics literature. They're easy to explain and once you become aware of them, you'll never succumb to their influence again—no more stereotyping, anchoring, ingroup bias, halo effect or succumbing to the not-invented-here bias. There's a lot to go through, should we get started?

Ha-ha, if only it were that easy! When I first lectured on an MBA programme, we used to cover about 15 biases in depth. I can't even remember how I chose those 15, but my students dutifully learned them and were able to correctly identify each in a case study, but were singularly unable to identify them in their own reasoning. It was just a theoretical exercise to them. Besides, biases can only be identified in others after information has been processed or a decision made, by which time it may be too late to change the outcome. Raising awareness of a few select biases doesn't hurt, but large-scale bias training, such as Starbucks did in 2018 across all their branches after several racial incidents marred their liberal reputation, has been seen by some commentators as only effective as a public relations exercise. Biases are shortcuts that reflect our ingrained values and beliefs, neither of which can be permanently altered by a 90-minute keynote encouraging us to do so. Such unconscious bias training is aimed at countering discrimination in the workplace and separates conscious from unconscious biases* that affect our views on age, race, sex, disability, religion or belief, gender reassignment, sexual orientation, marriage and civil partnership, pregnancy and maternity.

Research by the Equality and Human Rights Commission (March 2018) suggests that unconscious bias training 'can be effective for reducing implicit bias, but it is unlikely to eliminate it'. Interventions that aim to reduce explicit bias or openly prejudiced behaviour have also yielded mixed results because people 'tend to believe that they do not hold explicit prejudiced attitudes'.[12] This is why large organisations still implement hard controls, such as race and gender targets for hiring and promotion and other diversity and inclusion initiatives, rather than relying solely on soft nudges such as unconscious bias training.

So how, then, would *reading* about biases that could impact your information processing and thinking help you avoid them? It won't really. It will raise awareness, promote reflection and give you a set of labels or language to identify and discuss bias. I can't promise more than that. So, what do I propose then? Forget about bias and blunder on in our evolutionary imperfection? No, of course not. I'll take you through just a few biases that wreak the most havoc with our information processing and you can decide which ones might have crept into your thinking over the years. Then I will show you how to adapt your decision-making processes to neutralise specific biases. Cancelling out biases procedurally is far more effective, especially for teams, than relying on our ability to remember and identify them, especially when making decisions at speed.

---

* Unconscious (or implicit) biases are the views and opinions we hold that we are unaware of; they are automatically activated and operate outside conscious awareness, affecting our everyday behaviour and decision making. They are influenced by our background, culture, context and personal experiences. Training is more effective in reducing unconscious/implicit bias because it surfaces them, raising awareness, and so gives us some control over their impact.

The reality of being biased

"Cognitive biases are beyond the control of anyone. They originate from the depths of human nature and unconsciously distort the way in which we perceive the world and our own actions.**"**

—CEIS, 2016, Intelligence,
the Human Factor and Cognitive Biases

Let's start with a thought experiment. Imagine that you volunteer at a hospice (hospital for terminally ill patients). When you arrive on a Sunday afternoon, the nurse on duty briefly describes two patients and you get to choose which patient you want to spend time with. On one particular Sunday, these were your choices:

- Patient 1 can't feed himself, is incontinent, is emotional, can't speak coherently and bites at will. You'll need to feed him and perhaps wheel him around the garden.
- Patient 2 has memory loss, is frail, can feed herself, will want you to read to her.

Who would you choose? Unsurprisingly, almost everyone I ask chooses patient 2, with only a handful of respondents choosing the challenge of patient 1—some of whom had elderly parents themselves. Let me introduce you to the patients …

Patient 1 is a cute and chubby-cheeked two-year old, grinning from ear to ear as he chews on a plastic teething toy. He wears a nappy because he's two years old and not yet toilet trained, bites because his new teeth are hurting his gums and boy, can he throw a tantrum when he doesn't get what he wants. However, a quick stroll around the garden in his pushchair quickly settles him down again.

Not what you expected, right? You're not alone if you imagined a challenging old man with a somewhat disagreeable disposition. You probably imagined both patients in full colour without blanks in the picture to reflect holes in your knowledge. Your brain was able to generate these personas based on only the few words you read above. This is bias in action, although I do prefer to use the word heuristic rather than bias in such cases. A heuristic is a shortcut, and that's exactly what our brain took in, filling in the blanks and giving us the confidence to make a choice based on it.

What you just experienced was anchoring. Here we rely heavily on a few pieces of initial information and let that carry more weight than it should in our decision making. Which piece of information did you anchor on? It was, most likely, the word *hospice*. A home for terminally ill patients is not a typical place for a baby, but many a family has suffered this cruel fate.

Anchoring, as a mental heuristic, creeps into almost every situation where we have to weigh up information. What we already know about something is presented to us quickly and subconsciously and becomes our starting point, or anchor, for thinking about the subject. We tend to adjust and filter new information relative to our existing anchors.

Before my husband and I moved to Hong Kong from South Africa, we visited the island for two short days and hired a cheerful property agent to help us find a suitable apartment. She knew our budget and that it wasn't negotiable. She also knew that our knowledge of the local property market (and Asia in general) was limited. She showed us loads of apartments within our budget. They varied from mouldy to stinky, tiny and dilapidated and sometimes all of these together. This was unexpected, as we had thought our housing budget was reasonable. Clearly, we were wrong and asked her to show us some nicer apartments as we were running out of time. Of course, every nicer apartment was out of our budget. Did we pay up? Sure. Had we been played? Oh, yes. After moving to our overpriced shoebox, we discovered that there were wonderful apartments very much within our budget that she had chosen not to show us. She skilfully created false anchors for us to judge our spending power against.

Simply being aware that we have inbuilt anchors or that we are susceptible to being anchored really isn't that helpful. Starting any discussion or decision process with clarity on inbuilt beliefs, our anchors on the topic and imported anchors (data on which other parties are hoping to anchor you) can be very powerful. And not just because setting the opening anchor in any negotiations will frame the entire discussion, but rather because anchoring leads to a far more insidious, hard-to-spot mental shortcut—confirmation bias.

In March 2011, I was teaching a semester of Critical Thinking at a business school in Singapore. That month, a magnitude 9 earthquake off the coast of Tōhoku in Japan caused a tsunami that devastated the Fukushima Daiichi Nuclear Power Plant. This was the largest nuclear meltdown since Chernobyl. As more information became available about the genesis of the incident and subsequent emergency operations to contain rapid nuclear fallout, my students and I were able to evaluate the mental mistakes that decision makers were making under conditions of enormous pressure and international media scrutiny. Sitting in our lecture theatre after a hot lunch and shooting mental arrows at bad decisions made by others turned out to be quite easy. Even deciding how their decision making could be improved turned out to be easy-peasy.

But critical thinking in real time is not so easy or even intuitive. So, come exam time, I turned the spotlight on my students. Their final group assignment question pack included articles with emotive photos and information about both the human tragedy and technical and regulatory failures of the nuclear disaster at Fukushima. The final question was this one about nuclear power:

> Q: Considering the recent events in Japan, would it be reasonable to suggest that nuclear energy be phased out worldwide? Support your answer using any suitable resources.

What kinds of answers do you think the majority of these final year MBA students gave? None of them were nuclear specialists, so I wasn't looking for overly technical justifications for their decisions; instead, I was most interested in how they supported their conclusions in the absence of technical knowledge. The vast majority of groups replied that absolutely, nuclear energy should be phased out and they cited information related to the recent disaster almost exclusively. They chose to ignore the safety and efficiency of the majority of the world's nuclear power plants and anchored instead on the outliers that had gone wrong. They also ignored influential differences between geographic locations and government oversight that would affect the safety and stability of a nuclear power plant. *Recency bias* encouraged them to give almost 100 per cent weight to the most recent event, with less or no weight afforded to decades of data on the efficacy of nuclear energy. What I was looking for was an ability to consider recent events in Japan in the context of a wider nuclear power discussion—a much wider frame.

I repeated the case study and asked this same question of the following two postgrad and undergrad classes at the end of 2012—a year and half later. Do you think I received the same answers with the same justifications, despite using exactly the same material? Of course not. These students firmly believed that nuclear energy was beneficial and here to stay and quoted far wider sources of information.

What happened? In the aftermath of the disaster, the pain, suffering and enormous damage to life and property were spread across the news in blow-by-blow, by-the-minute updates. It would be almost superhuman to be unaffected by the emotions and human tragedy unfolding across Japan. To suggest that nuclear power was the way of the future at this point in time would have been emotionally challenging. Eighteen months later I received very different conclusions for the same case study. These answers reflected a good understanding of how human error had caused the devastation at Fukushima. It was also clear that nuclear power had regained public favour. What a difference a few months had made. It's not just my students that were caught up in this trap of vilifying nuclear power in the aftermath of the tsunami. Politicians across the world were called on to justify nuclear power capacity and investment. In Switzerland and Germany plans to extend or expand nuclear power plants were called off, despite neither of these countries having any geographic or regulatory similarities to those affecting Japan, or Fukushima, specifically.

The first group of students had fallen prey to a bias they knew very well in theory—confirmation bias. This sneaky mental shortcut grows like a weed from a mental anchor, which in this case, was the human disaster of the nuclear crisis. Despite already agreeing that the cause of the nuclear failure was manmade and not a consequence of the type of power plant in question, they still decided that it was bad for humankind. They found plenty of evidence to confirm their belief and ignored a growing body of evidence that disagreed with them.

As their lecturer, I think I failed here. What's the point of only being able to avoid biases in academic exercises but not when making real decisions, when it really counts? You'll see confirmation bias creep in whenever you have a particular opinion on something and attach greater importance to information that agrees with your opinion and discount information or people that disagree with you.

Now imagine you were passing through London's Spitalfields Market on your way to work one frosty winter morning. You notice a magician spinning a roulette wheel and handing out money to onlookers willing to engage in a gamble. You pause to look. You stop, and the magician puts £20 in your hand and tells you to look at it. You do. You hold it as he offers you a gamble with 50/50 odds, in which you get to win £50 or loose the £20 in your hands. You can also walk away, immediately, taking the £20 with you. Consider it a freebie. What would you do?

Yes, it would depend on your risk profile, but this was exactly the experiment that the BBC[13] conducted in a crowded market with a pile of £20 notes and a magician. Here some punters were presented with the immediate gain of £20. Most of these surprised recipients took the loot and walked on, not quite believing their luck.

Other passers-by were presented with a different scenario. Imagine now that the magician slowly counts out £50, placing each £10 carefully in your outstretched hands. You watch each note arriving in your palm like an eagerly anticipated plate of food placed on the table in front of you by a waiter. You close your hand on the £50, expecting to hear the magician say you can walk away with the cash if you so wish or stay and gamble. Instead he snatches back £30 of it and offers you a gamble (with 50/50 odds) to win it back. Or you can walk away with the remaining £20. What would you do? Might you gamble this time?

Your position in both of these gambles is identical: gamble for £50 or walk away with £20. Yet the response received in the second exercise of this hardly scientific study showed that punters were far more likely to gamble to win back the £30 that had been taken from them. The first scenario was presented as a gain of £20, and most recipients were happy to walk away clutching this unexpected gift. The second was presented as a loss of £30 and most punters went on to gamble the £20 they had in hand, hoping to win back the £30 that had been taken from them—even though none of the money was theirs to begin with.

This is prospect theory in action, and I've gone as far as I can in exploring heuristics without mentioning the founding fathers of behavioural economics—Daniel Kahneman and Amos Tversky, whom you first encountered in chapter 3. There we covered rationality in decision making in depth, and this rigid definition of rationality is what sparked them to think about all the ways that they disagreed with it. They disagreed with it because they saw every day how smart people made decisions that varied significantly from the gold standard of rationality. Kahneman had noted that utility theory represented gains and returns to gambles at absolute levels, yet he suspected that changes in gains and how they affected relative levels of utility might play a more significant role in choice behaviour. Tversky built on this insight by testing if we made different choices when faced with losses rather than just the gains found in utility theory.

The results were clear and staggering and became the foundation of prospect theory, for which Kahneman received a Nobel Prize in 2002 (Tversky had passed away by the time of the award). With this theory they demonstrated that humans are, generally speaking, risk seeking when faced with sure losses and risk averse when faced with sure gains (Kahneman & Tversky, 1979). We feel a loss more deeply than we feel a gain and so, 'when choosing between sure gains and gambles, people's desire to [gamble to] avoid loss exceeded their desire to secure [a sure] gain'.[14] This was shown to be a general quality of the human condition rather than something we reserved only for monetary gambles. They speculated that avoiding pain (from loss or other) at the cost of maximising gain was a useful survival tactic. Tversky and Kahneman's research also found that we weigh probabilities not with subjective utility but with emotion.

This is why political campaigns often focus on what we lose through immigration or being part of a trade or economic union, not what we gain. As we know these signals are far more powerful. Perhaps a population will take a risk on an unknown presidential candidate to avoid the sure losses that he says will come from open borders and multilateral trade agreements? You'll also see this with financial market traders, where they prefer to trade their way out of losses but are quick to bank their gains.

Sunk cost is an example of loss aversion in practice and not only in monetary terms. A client in advertising was wrestling with a decision to let a high-profile team member go. After taking my client through very much the same decision process I'm exploring with you, he had decided that exiting the employee was the right thing to do for all stakeholders concerned. He would consult their HR director and start the processes. A month later, he had done nothing about it and was visibly agitated by the state of affairs.

'What's bothering you?' I asked.

'Well, we've invested a lot in this woman, not only the extensive fast-track training, coaching and on-boarding that our shareholders have paid for, but I have worked extensively with her. I have invested a lot in her, I can't simply walk away from that.'

The time, financial and emotional costs that had been sunk into this bad apple meant that they were more likely to continue to gamble on her ability to improve rather than lock in their losses and move on. Once my client understood that this was an emotional response to loss rather than a reflection of the facts at hand (the bad apple is still bad despite all the investment in her), he was far more comfortable in executing his decision. Importantly, he didn't regret it later on.

What does any of this mean for your decision making? Well, it means that you are sensitive to how information is framed—just like the rest of us. Is information you are presented with framed as a loss or a gain? Would you trust a heart surgeon with a track record of one-in-five of his patients dying on the operating table (a 20 per cent failure rate) or would you prefer the surgeon with an 80 per cent success rate? The only difference here is framing.

Let's move on from gambling and heart surgeons to Einstein and dating to explore one last bias, for now.

In 1917, Einstein discovered a glitch in his new general theory of relativity. His equations suggested that the universe could not be in a stable state but was either expanding or contracting. Except that was impossible because the universe was static and stable. He quickly developed a workaround to satisfy this fact and return order to the galaxy. At first he called it the 'cosmological constant'—a constant introduced into his theory to 'hold back the effects of gravity' and maintain a static universe.

In 1927, a Belgian priest and astronomer, Georges Lemaître, cornered Einstein at a conference and pitched a theory that he had been working on, based partly on Einstein's own calculations. Einstein dismissed him out of hand, commenting that his calculations may be correct, but his physics was atrocious! Lemaître didn't give up and continued to develop his idea that the universe was indeed expanding, and that the cosmological constant was superfluous. It would take another 60 years for Lemaître's idea to be recognised as the dominant theory currently underlying our understanding of our universe—the Big Bang Theory.

Today most scientists accept that the universe originated from an incredibly dense, incredibly hot, single point in time and space and has been expanding ever since then. In case you were wondering, Einstein eventually warmed to this idea and later called the cosmological constant the biggest blunder of his career.

This introduces our final bias. Any idea what that might be?

Overconfidence

Overconfidence is often confused with someone simply having an overly inflated view of their abilities, knowledge, appearance, etc. Most people also assume there's a fine line between confidence and overconfidence. One can be confident in one's ability to perform a certain task because of the amount of practice, experience or preparation undertaken. If you know what you are doing because you have done it before, or the numbers really do back you up, or you've used a thorough decision-making process, then confidence is justified. When does confidence become overconfidence?

One official definition tells us that overconfidence occurs when our confidence in our ability is greater than our actual ability. Of course, this could apply to overestimating our actual ability, or our ability relative to others, as well as being overly certain that we are right in our beliefs. Apparently, most of us are overconfident in the intelligence of our romantic partners as well.[15] How is it possible that we are wilfully blind to such self-delusion? Surely life presents us with feedback that allows us to calibrate our beliefs to our abilities? I ask my students at the very first lecture of the term to write down what grade they think they will achieve in my class. I collect these papers and store them until the final grade is known.

Despite never having taken a 'Critical Thinking in Decision Making' programme nor being taught by me, the vast majority of them simply put down their average grade modified by how easy they think critical thinking is. It's a non-technical subject for the most part, so grades are inflated by at least 10–20 per cent. In addition, I seem rather friendly which, apparently, further inflates their estimate. So here we have both anchoring (on their previous average grade) and overconfidence that that grade will repeat, despite very little information about the class.

The British historian Arnold Toynbee reminds us that, as far as civilisations go, *nothing fails like success*. Could it be that we automatically expect past patterns to repeat themselves or past successes to be recreated in different contexts under different conditions?

There is no shortage of examples of overconfidence from all corners of political, corporate, social and academic life. Studies confirm its existence but also caution that culture and personality affect our susceptibility to the bias, as do wealth, gender and expertise. A bias so destructive that Daniel Kahneman himself earmarked it as 'the first he'd eliminate if he had a magic wand'.[16]

What have you been overconfident about? Wait, don't answer that yet. Overconfidence is a survival tactic and something that often gets rewarded, especially amongst academics, politicians, leaders and experts. This makes it even harder to identify and root out.

David Tuckett is Professor and Director of the Centre for the Study of Decision-Making Uncertainty at University College London and he reminds us that society rewards and sustains overconfidence: 'We would simply discard leaders who admitted they didn't know or gave us the truth, which is usually that the answer is filled with uncertainty.'[17] Part of decision making is trying to imagine what will happen in the future. We can't know this, and so must use information and how we feel about that information and the future to come up with plausible narratives. To get others to believe in our narratives, we need to seem confident and present compelling stories. Since working as a consultant to consultancies, I often find myself at odds with their time, cost and effectiveness estimates, for example: 'This project will completely transform your organisation, will cost only $x and will be completed in three months.' When it's clear to me that these goals are unobtainable (which is most of the time) I usually voice my concerns. While I never do this in front of the client, I have been threatened, removed from a project, told that I'm not a team player and that I have 'no faith' in my consulting team.

Being a realist has very little place in the world of consulting. But consultants aren't squarely to blame. Clients, too, suffer from ballooning timelines and budgets, yet they continue to award contracts to those consultants that are the most (over)confident in their abilities. 'There is a 60–70 per cent chance that this project will be successful' is a perfectly acceptable thing to claim, but I'd never work in London again if I admitted that to a client!

The real challenge here is that, outside of finance and advertising where executives can use psychology to profit from their customers' biases, in strategic decision making, executives must understand and counteract their own biases and those of their colleagues. There aren't many examples of this being done successfully, with recent research on the tongue-twisting *bias blind spot bias*[18] reminding us that most of us tend to perceive ourselves as less susceptible to biases than others.

I did specify right upfront that knowing about biases won't help you overcome them—it may not even help you identify them in yourself, only others. Have you just wasted the last 20 minutes reading this chapter and am I about to be fired as your author of the moment?

Before you decide, remember also that I promised to share a better way of approaching bias in decision making, so let's turn to that and you can judge if it could work for you or not.

Bias bashing in action
Below we will focus on debiasing protocols, using four real decision-making strategies built to improve team or organisational decision making.

I don't know what biases *you* suffer from, I'm barely certain of the biases I suffer from. So rather than a five-step plan to live bias free, I'll share some examples of team decision strategies developed to reduce bias, as well as relaying some advice to straighten out individual decision making from experts in this field. Although I am less convinced of the effectiveness of the latter, as it involves us genuinely wanting to live bias free—which goes against our inbuilt survival and thinking habits. Consider it a challenge.

Here are some examples of debiasing strategies for teams:

1. Investment team maintaining their success

The B Team was a very successful investment team based in Cambridge. The chief investment officer knew this; he also knew that his team was fairly inexperienced and, at the time of working with them, only two team members (the old hands) had experienced a significant and prolonged negative market event, such as the tech bubble bursting and subprime housing crisis. Their investment process had been built over time in fairly benign market conditions with a team that worked exceedingly well together. He worried that they weren't sufficiently positioned for a market downturn or that, perhaps, they were getting too comfortable with each other's views, which resulted in a lack of robust challenge. After a morning of observing them tackle several decision-making challenges, I was able to see that the old hands considered themselves as the team guardians and guides and so had a larger share of the team's voice than the others, despite insisting that everyone was heard equally. Others waited for them to take the lead and deferred very naturally to their experience. There's nothing wrong with that if you're training a new team and the old hands take responsibility for the team's performance; however, the B Team was a mature team where everyone was expected to step up to increased responsibilities. The CIO worried that their desire to maintain convivial working relationships was preventing them from truly challenging each other's ideas, which would, eventually, impact their investment performance.

Together we explored changes that occur in the dopamine pathways in our brain when we collaborate as a team. How feeling part of a close-knit team reduces stress and increases feelings of well-being and how this results in the desire to maintain team harmony. We also explored how disagreement and direct challenge from our team-mates travels along the same neurological pathways as physical pain, which provides a handy explanation for why we tend to avoid giving and receiving challenges. My experience in working with British teams is that they live up to the stereotype of being polite at the expense of effective challenging and robust decision-making. This is a trait that I found in Japanese and Singaporean teams, too, but for slightly different cultural reasons.

With little experience of effectively challenging one another and carrying concealed cultural baggage that equated challenge with aggression, this team had to learn how to challenge ideas effectively and so to trust each other even more. They had to learn to trust that a challenge was out of loyalty to the process and the collective outcomes for themselves and their clients. It wasn't personal.

Why not simply designate a devil's advocate at each meeting? In fact, this is never a good idea because of the toll that it inflicts on an individual. It's physically taxing to challenge others, even if it's specifically asked of you. If you are not naturally inclined to pull someone's argument apart, it can create mental dissonance, increase stress levels and feelings of not belonging in your team. I've seen how designated 'devils' look for ways to challenge with the lightest touch, so as to minimise the impact on themselves and their relationships. It simply doesn't work in the long run. Instead of dropping in a devil's advocate, the B Team formalised their decision-making process with decision-making protocols.

An example of a decision-making protocol for investment teams would be: instead of only presenting one investment idea at a time, each analyst is asked to present two at the same time so as to reduce the potential of sunk cost effect, loss aversion and ownership bias. Investing time and energy researching an investment option creates a sense of attachment. The more time a team spends on an idea, the harder it becomes to see its flaws and create a realistic view of the opportunity. Once we've bought a stock and so formalised our relationship with it, it becomes harder to sell, regardless of whether it performs well or poorly. If it performs poorly, selling is admitting we were wrong (remember we are loss averse and prefer to gamble with losses than lock them in). If it does well, severing the relationship is a gamble that it will not continue to reward us, and we prefer not to gamble with gains. It's a complicated relationship.

The team as a whole can then decide which idea to explore further. The analyst that proposed it is tasked with presenting both the merits and demerits of the investment. This allows a robust debate within the team about the negative information that the analyst presented. In this situation, both agreement and disagreement are positive, allowing the entire team to form a balanced view of the investment opportunity rather than see it through the rosy lens of the analyst advocating for it.

Of course, this strategy would work with most investments or projects.

2. Venture capital firm defining their edge

The D Team belonged to a venture capital firm, with a medium-to high-risk profile nearing the top of their game. But they hadn't always been there. For a venture capital firm to find and invest in the next massively disruptive technology or product on terms that would compensate them for the risk they would be taking on, they must be sure of picking a winner. Of course, this kind of certainty isn't possible. It does help if other firms see the potential in a new technology or product, too. This makes an investor feel a little more comfortable about their bet. Yet, if others also see the opportunity, the price goes up with demand and returns decrease. One strategy that helps overcome the safety of the crowd or herd instinct is a contrarian one.

This particular venture capital firm wanted to ensure that the technologies they invested in were truly rare and undiscovered gems. If everyone on their five-person investment team agreed on a particular investment, then they would throw it out! Such a clear winner wouldn't suit their risk profile; plus, if they could all see the potential, then so could every other venture capital firm out there. Instead, their decision-making process correctly acknowledged that consensus-driven decision making reflects a low risk/low return strategy. To counteract this, they would only invest in opportunities that were rejected by at least 30 per cent of their team and accepted by 60 per cent. Naturally, this gave absolute freedom to team members to fully explore the risks in an investment and debate the downsides, knowing that such disagreement would strengthen their understanding of an opportunity. I will add that this is a mature US-based team, very comfortable with robust debate and with a good understanding of each other's personal risk preferences and biases.

3. Changing the face of ethics in Big Pharma

Big Pharma has a bad name. Sure, these companies bring us life-saving drugs and continuously invest in research to find new cures for debilitating conditions, but the price they charge for these is often seen as excessive, unrealistic and unethical. As greed gobbles up the headlines, some pharma's are quietly working to show that not all of them put profit before patients. Today, all Big Pharmas have extensive codes of ethics and patient-centred practices, so standing out in this area is a challenge. Company P is a global pharmaceutical wanting to change negative perceptions in the market by making a tangible difference in the way they work, from the top down and the inside out.

Every company is a decision-making factory, and these decisions determine not only the success of the organisation but public perception of it. They knew that they already had good people working for them and a solid ethics programme, so decision making was where they focused. What could they give their teams to help them make the most ethical decisions they could? Firstly, they stopped using the term 'doing the right thing'. As a large multinational, doing 'the right thing' in a particular circumstance to an employee in China could be different to what an employee in Brazil might think is the 'right thing' to do in similar circumstances. What biases did almost 10,000 employees harbour, and how could they avoid these in decision making?

Values are core biases that differ vastly across nationalities, generations and individuals, so their ethical decision-making process began by asking all employees to get clear on their own values and identify where these weren't aligned with clear corporate values. Leaders were trained to help employees do this and so were their newly trained ethical decision-making coaches. They proceeded to build a decision-making strategy that would help employees identify an ethical dilemma, ensure they are solving the correct problem, identify stakeholders and explore impacts on those stakeholders over time. Requesting opinions on a decision from a superior, a colleague with a different viewpoint and a third colleague with no stake in the decision was also part of the process. Together, these steps allow decision makers to thoroughly explore the problem, test their assumptions and agree on stakeholders and current and future impacts—all while rooting their discussion in organisational values. The real power of this framework was that it allowed them to call out assumptions that different team members could unknowingly bring to the problem domain.

4. Supplementing fast and frugal decision making

The final example is X Inc., one of the world's largest manufacturing organisations. They had a mandated, fast-and-frugal decision-making process that allowed them to keep up with a rapidly changing consumer environment. It was all about being agile and disruptive, overcoming deadlocks and widening the bandwidth of their decision makers to ensure they weren't bottlenecks in a VUCA environment. I'm not a fan of speed at any cost (nor business jargon) and the cost of this strategy was that decision makers did not have to take the time to explain their decisions to subordinates. This eroded trust between colleagues, as decisions could be taken unilaterally with little consultation. They also didn't need to speak to people they knew would challenge them and so slow them down, even if these people were the experts. Unsurprisingly, disempowerment and disenchantment ensued.

On the flip side, many employees didn't have the confidence to make decisions quickly or unilaterally and so escalated, procrastinated or avoided making a choice. The process was broken—but implementing a long, detailed decision-making framework was never going to appeal to them. What was the problem they needed to solve? Firstly, they all needed to be aligned about the future of the organisation so that decision makers always had the same ultimate goal and could justify their choices in line with it. Rebuilding trust would come through openness and consultation within the decision-making process. A common language for discussing decision making and approach to risk would also be helpful. Their fast-and-frugal system was augmented with specific steps to slow them down and introduce essential critical thought and challenge into their thinking. We agreed on a simple framework that looked like this:

1. Resolve decision rights—who is the decision maker?
2. Check you are solving the right problem.
3. Call out frames.
4. Make assumptions and risks explicit.
5. Gather two alternative opinions.

You can see from the above that the meta-decision, and identifying biases, blindspots and past experiences have all been left out of this framework. A simple five-step framework worked with their fast-paced culture—anything more would have turned staff off.

This framework helped decision makers feel more confident in their decision making, mostly because their ultimate goal changed from making *quick* decisions to *effective* decisions. A more robust framework also ensured that they examined their decisions from more angles, better understood the pitfalls in the information they gathered and were better able to articulate their choice rationale through the lens of the framework. An important step in the process was gathering challenging opinions so that this morphed from something to be avoided to something to actively seek out— and, in so doing, it became easier to give and receive scrutiny.

Sharing specific terms to use when exploring risks, assumptions and emotions with their teams made it easier for them to think systematically and have constructive conversations about these. Did it take them longer to decide? Yes. Did the business suffer for it? No.

What all of these examples share is a willingness to adopt a process orientation in their decision making. They didn't try to identify and eliminate individual biases amongst staff, but rather focused on what outcomes were important to them—and then went in search of those behaviours and biases that would be the most damaging to their decisions, and hence, success as a team or organisation. You might recognise their first steps as solving the right problem.

What about you as an individual decision maker? What biases do you think plague your decision making? What biases should your decision-making process counteract? This is a really difficult question to answer and requires some honesty, reflection and dredging up memories of decisions that didn't go so well—and who likes to do that?

PART 3

11

RISK AND UNCERTAINTY

How your risk persona taints your view. Telling
tales about risk to understand them better and
tools to root out assumptions.

Here's a quick recap of where we are in our best practice decision-making process. Remember that decision processes are highly personal, but many include at least three or more of the following success factors:

- a process rather than outcome orientation
- clear decision rights
- a meta-decision (including ensuring that the correct problem is being solved)
- checking that the problem is being correctly framed
- being aware of the mental biases you/your team are most prone to
- exploring assumptions and risk ◄— *we are here*
- understanding the role of unconscious processes on risk perception
- gathering challenging opinions

Risk persona as a lens

All decisions involve risk—idiosyncratic risk from the decision maker, quantifiable risks, risks from assumptions and what we don't know, we don't know, alongside black swans and snakes in the grass. Booms, busts, bank runs and corporate failures are part and parcel of our complex and risky political, financial and social environments. The risks that drive extreme events are often the ones that no one paid attention to or could have foreseen when making the decision or setting policy. While it's very hard to know what you don't know, the ability to imagine alternative futures is becoming more important around the boardroom or WeWork table.

Because risk identification and management are so important in the business world, we like to leave all things to do with risk up to the risk manager and his or her flock of actuaries and PhDs. Unfortunately, this behaviour is in itself risky business. If individual decision makers don't have a strategy beyond models and numbers to grapple with unprecedented risks and imagine the unimaginable, then unimaginable things will continue to blindside us. The subprime crisis that swept global markets in 2007/2008 has largely been labelled a crisis of imagination, where politicians and governments alike failed to imagine that such an outcome was possible and later admitted so. I try not to use the word 'imagine' with my corporate clients, but no other word seems to fit the bill as snugly. Their risk processes failed to flag the risk of global systemic failure or the possibility of a bank run because the risk systems used had been programmed by minds that couldn't—or wouldn't—imagine such an extreme financial event.

"Problems with individual financial sectors
were identified, but a global failure of imagination
meant no one anticipated this crisis.
No one stopped to think 'what if'."

—Michael Coogan, Director General of the Council of
Mortgage Lenders

The lens of the past and present curtails our ability to imagine the unimaginable and so many states of the world remain unknowable, filling our decisions with uncertainty that we can't measure, assign probabilities to or manage. Risk, on the other hand, is more measurable and manageable—in theory. This requires that we assign possible outcomes and the probabilities of each occurring to the available options within our choice analysis. Yet, in reality, alongside these quantifiable risks, many of our decisions involve states of the world that we cannot foresee. We simply don't know what we don't know.

Any possibility that is assigned to a probable outcome in an unknowable future is subjective and would require forecasts and speculative narratives to sustain it. But before we get too disheartened by uncertainty's confidence-busting powers, let's explore one category of risk that we have some control over— ourselves. More specifically, our own risk tolerance. To be clear, personal risk tolerance is a frame, or bias, that affects how we think about risk.

I have pretty much always been risk averse. In fact, I'm so risk averse that I used to struggle making any decisions at all, because I became paralysed by the idea that if I chose one option then all the other options were no longer available. Yes, clothes or gift shopping with me is a tedious ordeal as I circuitously debate the merits of both buying and not buying an item of insignificant value. It took me eight years to complete my undergraduate degree, not because I am particularly slow or unfocused, but rather because, without any career guidance, I changed my major subject three times. Every time I settled on a subject, I faced debilitating regret—not that it might prove to be a bad choice for me but rather that the other options might have been better. The choice of a partner and career were also offered at a time when I didn't have a mental database of decision-making outcomes to learn from. In both deciding to get married and choosing a career, the same decision paralysis endlessly overshadowed my thinking. In the former, I sought a sounding board in the form of a psychologist to help me decide. I eventually said 'yes' and, after some 20 years of a fulfilling marriage, I am very glad I did. In my career, I simply went with what was on offer, going with the flow, or deciding not to decide.

It was this debilitating uncertainty that led me to a career in decision making, where I have learned to promote my unconscious risk aversion to a conscious fear of taking risk. It's put me in control. This career has allowed me to explore the effects of being hypersensitive to risks and plain old risk averse in my thinking. It is a trait gleefully exploited by insurers and sales tactics of all kinds. I know that I pay a premium to wrap my life in a protective blanket of insurances, from super comprehensive car insurance to insuring theatre tickets and even summer camps against the possibility that my son falls ill. At best, most of these are not necessary and at worst, statistically foolish. When my husband and I explore new opportunities for our family, like moving to a different country (we have lived in four countries, six cities and 11 different houses), or buying a holiday home or choosing a school for my son, our views of what is important are vastly different because we have risk personas on either end of the spectrum. After 20 years of being together we use these perspectives to strengthen our conversations and temper our most extreme ideas. Despite knowing that I have a warped lens through which I view risks, I have been unable to change it. Instead, I have learned to adapt to it—to counteract its effects at times and simply accept it at others.

Yet, according to Daniel Kahneman,[1] *being less certain* is the first rule of good decision making—*being less certain about everything!* Here, Kahneman means for *uncertainty* to be used as a tool to counteract overconfidence and assumption. If you are certain that choice A will lead to outcome Y, the only way to be sure that you're not overconfident is to test the logic of your thinking. This requires doubt and curiosity. We can never really be certain of anything, but we are allowed to have confidence in our reasoning. My journey into understanding decision making was a journey of gaining confidence in my thinking and allowing my doubt to serve as an indicator of where I need to pay more attention or gather more information or opinions.

Today, I am more comfortable in deciding because I strive to understand (as comprehensively as I can) what I am giving up as well as gaining in the choices that I make. There is always risk from the things I know I don't know, the things I don't know I don't know and the things I can't control. Of course, I still can't really quantify uncertainty, yet somehow, this little charade (and a decent decision-making process) makes it easier to be confident in my choices. But just how confident should I be?

For Daniel Kahneman, overconfidence is the greatest threat to our decision making. Being circumspect and cautious is fine when you are an academic like Kahneman but erring, cautious, risk averse leaders don't really inspire much confidence. Confident leaders aren't better at forecasting the future compared to the rest of us and their decisions play out in an uncertain future, just like ours. What they do better is create narratives around what they know, extrapolate this into the future and so create a believable alternative, complete with known risks and risk mitigating strategies. Inspiring leaders, investors and headmasters are all master storytellers whose confidence is contagious. To maintain *our* confidence over time, at least some of their stories must become reality—and so a storyteller, like a research scientist, will head out to prove their theories.

It would seem a bit pointless to disprove your own theory. Yet there is an important place for disproving theories and narratives in good decision making, too. Attempting to disprove our own ideas gives us some ammunition against anchoring, confirmation bias and *believing-one's-own-BS bias*. Ok, I made the last one up as a snappy synonym for overconfidence. This can be done by asking one team member to come up with reasons to continue to support a project while another finds reasons to abandon it. For example: a (brave) CEO could task a team with highlighting the vulnerabilities and assumptions in his five-year strategy plan. An FBI team investigating a high-profile, highly-contested case could split into two teams: one team to gather evidence of innocence and another to gather proof of guilt.

Once we come to terms with our tendency to be overconfident and fall in love with our stories and the world they predict, Kahneman[2] suggests that we test what we think we know with some hard data and so revisit the logic of our decision.

Perhaps asking what other options would you have if you were less sure that A would cause Y, or that Y is preferable to X? Have you considered a dramatically different outcome to your preferred one? What are the assumptions that underlie your choice?

Base rate realities

Questioning the base rate or assumptions that you use as your anchor is another piece of advice widely offered by behavioural economists and statisticians. For example: if you believe that people who work for themselves make more money and are happier than those with a 9-to-5 corporate job and this belief drives you to fantasise about setting up on your own, you should probably test your assumptions before you resign your current job. How many professionals are in your field and how many are freelancers? Is that number growing or falling? What do they earn on average? How have earnings changed over time? How many years does it take to establish oneself as a freelancer? Are they happier or more stressed about their earnings? Do your ideal clients have minimum requirements for the size of an organisation that they will contract with? Does it include freelancers?

Or imagine you have a malady and are offered a particular medication with an impressive track record of curing over 10,000 people. That sounds good, but we can really only judge it if we know the number of people that have been treated. If 100,000 people have been treated and only 10,000 have been cured, then it doesn't look so good. Questioning base rates built into statistics is a good way to identify assumptions.

Decision making usually begins with an inside-out process of understanding the decision, our options and their impact on us. Starting with base rates and statistics instead allows us to start with an outside-in view, which has a wider frame and a less personal lens. But base rate assumptions are the blandest kind of assumptions— let's take a look at some other varieties of assumptions.

Assumptions: from fiction to facts

A global packaging company is a client of mine. I have learned far more about packaging and plastics than I ever thought I would. But it wasn't always that way. As a complete outsider, I have been able to flag some of the assumptions that are so implicit in their thinking, they don't ever stop to question them. In working on their decision making around sustainability, certain things were simply taken for granted, like the 'fact' that plastic bottles needed plastic labels and plastic lids (secondary plastics are a significant polluter) along with the 'fact' that there would always be a market for plastic-packed drinks and that the next generation of their enormous global following would stay as loyal to their brand as current consumers, despite changing tastes and values. By calling out these assumptions and challenging them, future risks can be clearly highlighted and seeds for innovation sown.

Assumptions play a starring role in our decisions. Sometimes born from the status quo and sometimes from what we don't know or beliefs we haven't updated. They are stories we create and repeat so often that they sometimes graduate to 'facts' in our thinking.

A few simple questions help to expose the assumptions hiding within our reasoning but, as always, these are only effective with a little bit of honesty.

- What do I/we know but can't prove?
- What do we accept without challenging? What is the status quo?
- What do we not know?

The last one is not a trick question. It refers to our ability to make stuff up to fill in the gaps in our knowledge; to tell tall tales that carry the veneer of respectability and could even be assumed to be true. If we are able to acknowledge that we truly don't know the facts of a matter or how something will turn out—or where interest rates will be in two years' time—then the quality of our conversations improves immediately.

There are many challenges in bringing a new aircraft to market. The Boeing 737 Max was almost the perfect example of radical innovation, featuring a cockpit and controls pilots were already familiar with, requiring little adjustment or training to be able to fly it. Despite its familiarity, it had undergone an internal overhaul to improve efficiency. Within three years of its maiden flight, 346 of its passengers had been killed in two separate and avoidable crashes. The blame was squarely rooted in these internal changes; yet, on closer inspection, the machine wasn't to blame, but rather the same two assumptions made by its engineers, test pilots and the FAA (Federal Aviation Authority). These assumptions were brought to light by the *New York Times* in June 2019:[3]

"A year before the plane was finished, Boeing made the system more aggressive and riskier. While the original version relied on data from at least two types of sensors, the final version used just one, leaving the system without a critical safeguard. In both doomed flights, pilots struggled as a single damaged sensor sent the planes into irrecoverable nose-dives within minutes, killing 346 people and prompting regulators around the world to ground the Max."

These sensors are placed near the nose of the plane and are often damaged by bird strikes and jetties (moveable stairs) bashing against them. Up until now, two sensors were able to feed information into internal systems to compensate for any loss or damage to one. Naturally, Boeing employees and those at the FAA assumed that the system continued to rely on data from more than one sensor. They also believed that this system would rarely, if ever, activate, which is why the effects of a faulty sensor were never tested. These two assumptions were the basis of many critical decisions concerning design, certification and training.

"'It doesn't make any sense,' said a former test pilot who worked on the Max. 'I wish I had the full story.'"

—*New York Times,* 1 June 2019

With a finely tuned ear you will find assumptions all around you. But I would advise you to only root out those assumptions that impact your decisions and well-being rather than every stray assumption that crosses your path. Personal experience has taught me that this is not the route to popularity.

States of the world

Do you remember the definition of a rational decision maker from chapter 3? Probably not, so let's recap; in theory, a rational decision maker is one who selects the option that would maximise their expected utility (pleasure or gain) for a particular level of risk. For this theory to be useful, the decision maker has to know what the probabilities are of outcomes occurring: for example, if you are playing a board game then the probabilities of a fair dice landing on each number from 1 to 6 are known. However, in everyday decision making we don't always have the luxury of such objective probabilities.

In 1954, Leonard Savage extended expected utility theory to circumstances in which the actual probabilities of outcomes occurring were unknown. This subjective expected utility theory (SEU) allowed for probabilities that were influenced by assumptions and beliefs of the decision maker. For example: if you were securing a mortgage to buy a new house, you could work out how each level of interest rate impacted your payments and hence cash flow. You would also need to have a view on how likely each of these interest rate levels were. This would be based on your (or your advisor's) view of the economy over the life of the mortgage. No one can know how an economy will perform over 20 years, so these probabilities would be a belief-fuelled guess. However, you would have been able to test your ability to weather extreme economic events and decide if you wanted to take the risk if you knew you couldn't meet payments at high levels for prolonged periods.

This allows you to distinguish what is under your control and what isn't as a decision maker and thereby quantify your uncertainty.

The interest rate is an external variable over which you have no control and so it should count as a random variable within your decision. Your ability to generate income and other future financial commitments are within your control but uncertain, and so can be categorised as uncertainties.

Once this distinction is clear, modern decision theory offers a nifty suggestion to help you in your choice: divide a decision scenario into *actions, states of the world* and *outcomes*. States of the world are possible scenarios that may unfold and create a unique outcome with unique impacts that the decision maker can't control. Actions represent the decision maker's feasible choices. These interact with each state of the world to produce unique outcomes that can be mapped in a matrix. These can be ranked along a utility curve (a list ordered by preference).

	State of the world 1	State of the world 2
Action 1	Outcome 1	Outcome 2
Action 2	Outcome 3	Outcome 4

Action 1 could be buying a house well within your budget, but one that you will have to extend as the family grows. Action 2 could be taking a bigger bond than you planned for to buy the house of your dreams now, which will be big enough for the family you are planning to have someday.

States of the world might reflect the interest rate you could possibly pay at different levels, the state of the property market in five years' time, or whether you started your own business or not and how that affected your income (the outcome). The real magic is taking a guess at how likely each of these outcomes are and whether you are able to weather the associated risk. While there are no absolutely correct answers here, how you rate and rank these likelihoods or probabilities will depend on your own risk tolerance.

Quantifying every possible decision state and every possible influence that will impact the outcome is quite a tall order, especially since some of these are unknowable or occur in a theoretically infinite set. If we really can't make fully informed decisions about the future, should we rather just go with our gut— which, after all, has been shown to be much better in practice than in psychological experiments?[4]

Not entirely. As problems with calculable answers get relegated to AI and machinery to sort and solve, the challenges that remain are those that don't have one right answer. Executives will increasingly only deal with those issues loaded with ambiguity and uncertainty beyond what an algorithm can digest. This is where our ability to wrestle with the unknown, to tell tales and translate risk appetite into preferences and perspectives, will allow us to deal with an increasingly complex environment.

In the next chapter, we'll explore how gut feel and emotions both help and hinder our thinking. Needless to say, in this state of the world, gut feel as a standalone decision-making tool is usually inadequate. A range of tools and theories to challenge our data, our thinking, assumptions, risk perception and mental shortcuts is no longer optional in a successful career, or even life in general.

(12)

WE ARE ALL
BRANDON MAYFIELD
(CASE STUDY)

Meet the good guy, the bad guy and the Feds.
Grade the FBI using what you've learned so far.

Three a.m.: he lay awake again. The same noises shuffled in from the lawn; formless noises like footsteps and whispers through softly rustling trees. Every day that week on his way to work, he had noticed the same bearded man standing on the corner near the station café. With eyes hidden behind dark sunglasses, the man seemed to track his progress across the platform. He wondered if he was being followed as he passed the newsstand where headlines flashed the trauma of at least 200 people killed and thousands injured in a terrorist bombing in Madrid—Spain's own 9/11. At first, he thought he was simply being paranoid. Had helping a Taliban supporter in a custody battle over a year ago come back to haunt him?

Four a.m.: more muffled voices in the yard. Every day he grew more certain that he was being watched. Doors that were open when his wife left home were locked when she returned. Small things were out of place in the house; a hairbrush in his bathroom, a clock in the living room. He prayed to Allah that his family would be safe. After all, he had done nothing wrong and had nothing to fear.

This is the position Brandon Mayfield was in on 6 May 2004, when two FBI agents knocked on the door of his law office in Portland, Oregon. At first they were polite, just wanting to ask him a few questions; but as they eventually forced their way through the door under cover of a warrant for his arrest, Brandon knew something more serious was going on—really serious. He soon learned that his fingerprint had been found on a bag of detonators near the scene of the terrorist bombings in Spain. The print was a near perfect match.

Mayfield had converted to Islam to marry his Egyptian wife, Mona. He had defended a convicted terrorist in a child custody case, giving him known links to a terrorist organisation, plus his computer held further 'evidence' against him in the form of searches for 'flights to Spain' and 'flying lessons'. In addition, he'd received combat training during his time in the US military and his fingerprints were on file. Surely Mayfield was a terrorist and the FBI had their man?

Brandon Mayfield was a lawyer and a father of four who gave his time and expertise to help those who couldn't afford high legal fees. His passport had long since expired and he hadn't left the United States in over 10 years. Despite this, the FBI had a solid case against him, with only the opinion of an independent forensic expert outstanding. This expert was brought in to verify the FBI's findings and his verdict would largely seal Mayfield's fate. Surely this man would find the error in the FBI's analysis, surely this was all a huge mistake? Unfortunately, the forensic expert also decreed with certainty that the print on the bag in Spain was indeed that of the accused. '100 per cent Verified' declared the FBI report. The full weight of the justice system now pinned Mayfield to the wall, yet he maintained he was wrongfully accused.

It is worth noting that FBI fingerprint examiners are governed by the 'one discrepancy rule' in which a single difference between a latent print (that being examined) and a known fingerprint must result in a non-match being declared, unless the examiner has a good reason for not doing so. Latent fingerprint identifications are subject to a standard of 100 per cent certainty. And so it was that four different examiners had declared with 100 per cent certainty that there was not a single discrepancy between the prints on the detonator bag and Mayfield's prints.

The Spanish government wasn't so sure. After Mayfield had been held captive for two weeks, it emerged that the Spanish police had repeatedly informed the FBI that they had found another match for the print on the detonator bag—an Algerian national. Their suspect had a credible motive and had actually been in Spain at the time. Given this new information, the FBI was obliged to release Mayfield. How could the FBI have been so wrong?

Former CIA employee, contractor to the NSA and whistle-blower, Edward Snowden, has publicly spoken out against the dangers of mass surveillance, for both ethical but also practical reasons. An enormous pond of stagnant data (now called a data lake) allows anyone with the necessary clearance to retroactively mine pieces of information, pick out the ones that confirm the theory of the day and create a plausible story to back it up. I use the term 'clearance' rather lightly as we now know that our data is gathered en mass and used liberally by service providers. Anyone with enough data can reshape the puzzle piece to fit the hole in the puzzle or create a tangential chapter in our story.

The Mayfield case is particularly striking because of the number of 'experts' convinced of his guilt, despite the lack of any concrete evidence. It seems that there was enough information available for these officials to pick the facts that fitted the case against him. They suffered from a rather public case of confirmation bias, aided and abetted by the number of 'facts' on the table. The results showed that their thinking tools weren't as sharp and well calibrated as they should have been. Thankfully, the Spanish police acted as a counterbalance in this case, but what about the many cases where the decision of a single agency, group or even person is relied upon?

Just from this brief scenario, can you find reasons that support Mayfield's guilt?

Looking at the case through the frames of 'Can I prove him guilty?' or 'Can I prove him innocent?' changes the information that you pay attention to, such as whether you focus on similarities or differences between fingerprints. If you now look for reasons that support his innocence, you'll come up with a different list.

An official review of the handling of the Mayfield case[5] was launched in 2006 with findings published in a 330-page report. It indicated that the FBI gathered and evaluated evidence through the single frame of proving him guilty and so *ignored contradictory evidence*. Such a high-profile case was handed to three FBI forensic examiners in the first instant and so any questioning of their judgement by an external examiner would be hard to substantiate and, possibly, professionally unappetising.

How might a meta-decision have helped them get their ducks in a row before they began their investigation? Asking 'What problem are we solving?' and 'How will we go about doing it?' would have been helpful. If they had an opportunity to articulate the intention behind their problem solving, it might have sounded something like: 'We want to nail the terrorist who did this as quickly as possible'. Perhaps, after the first fingerprint match was known, they would have heard themselves say: 'Now we know who did this, let's go get the sucker!' I confess, my caricature of the FBI officers here is tainted by cliché, but you get the idea. More helpfully, they could have investigated several lines of enquiry and gathered evidence to prove both guilt and innocence—especially in such a high-profile case. The Spanish police informed the FBI laboratory that it had reached a *negativo* (negative) conclusion with respect to matching the latent fingerprints to Mayfield's, yet this too was ignored and the investigators missed an opportunity to catch their error.[6]

What other mental biases can you spot in this story?

Overconfidence manifested in various guises, from the FBI laboratory's overconfidence in the superiority of its database and examiners[7] and their ability to create convincing stories to explain away irregularities in fingerprints.

Anchoring and confirmation bias, where the fingerprint verification was 'tainted' by knowledge of the initial examiner's conclusion (anchoring) and by Mayfield's representation of a convicted terrorist and other facts developed during the field investigation, including his religion, Islam (confirmation bias). These biases also likely contributed to the examiners' failure to sufficiently reconsider the identification after legitimate questions about its accuracy were raised.

A forensic agent on the case summed up the framing effect by adding that, 'If he had been someone else, with a different history, we wouldn't even have considered him.'[8]

13

UNCONSCIOUS PROCESSES

Emotions don't need us, but we need them: a cocktail menu
of emotions and how they influence our risk perception
through incidental, specific and carry-over emotions. Plus,
the role we can play in exploiting and managing them.

As far back as 1872 Darwin observed the intoxicating power of
emotions to drive actions, superseding both will and reason.[9] By
the 1980s it was proposed[10] that emotions could not only operate
independently of cognitive thought but also in advance of it.
More recent imaging studies[11] confirm that emotions can indeed
operate independently or in anticipation of conscious thought.
Unfortunately, saying or doing something without thinking
about it is still not a legitimate excuse for bad behaviour! As soon
as we become aware of our emotions we should, in theory, be
able to act on or counteract them. The more we know about our
emotions' potential to influence our decisions, either consciously
or unconsciously, the better we are able to stay in charge.

Research in the area has exploded over the last decade, but I've
found some ideas to be more useful in practice than in theory.
Ideas such as: understanding the origins of the emotions that
influence our thinking; how specific emotions influence our risk

perception; where gut feel comes from and what role it should play in our decisions; and finally, do men and women really think differently and what impact could this have on our decision making as individuals and in teams? As you can see, this is an area that fascinates me and has been incredibly useful, both in my personal and professional decision making, not to mention the decisions made by my clients around the world.

Emotions as a source of risk: understanding the origins of the emotions that influence our thinking

Emotion is the language of the body. It tells you and others how it is feeling, it intercepts incoming communications and issues instructions that take evolutionary precedent over other forms of communication. It is quicker than reasoning and outpaces speaking, yet can't tell the difference between what's real and what's imaginary. You can read a novel or watch a 30-second YouTube clip and experience the complete gamut of emotions, and even have those emotions affect your behaviour for a while afterwards. It is this ability to respond indiscriminately to incoming stimuli regardless of whether it is 'real' or not that allows us to explore the true impact of emotions on our decision making in a simulated setting.

In an ongoing client engagement, my team and I created two competing companies as proxy organisations for participants to inhabit and make decisions in. Every detail was created, from corporate ethos to branding, financial statements, CSR missions, client profiles, ad campaigns, supply chains and legal and compliance policies. It felt authentic and 500 participants inhabited these organisations for half a day, rising to the decision-making challenges that we set each department as we observed their decision making in action. It wasn't real, yet the frustrations, misunderstandings, lack of communication and assumptions that developed as they interacted produced exactly the same emotions one would find in organisations today. Intense, real emotions compacted into a short space of time. There are several ways through which emotions impact our decision making and several themes[12] surfaced in this exercise.

Integral emotions influence decision making

Integral emotions are those that arise from the situation itself. For example, in the above scenario, placing non-financial people in the fictional finance departments immediately affects how they feel about the challenge and the data presented. Does being outside of their comfort zone push them to make riskier or more conservative decisions? Unsurprisingly, the more insecure someone feels in their role, the less risk (on average) they take on in their decision making.

These integral emotions arising from the choice at hand, strongly and routinely shape decision making.[13]

We also noticed that if someone is required to work with a colleague they don't particularly like, their team's decision making slows down. Feeling overwhelmed by too much information, or too little time to process it all, has similar effects. These emotional impacts occur subconsciously and arise purely in response to a particular situation. Being in a stressful environment, such as preparing for an exam or going through a divorce, will also fundamentally impact how one decides.

Incidental emotions influence decision making

Not only are you buffeted by emotions that arise from within a situation, but you are also affected by emotions that carry over from previous situations. If left unchecked, such an emotional hangover can cloud your judgement for as long as those chemicals course through your body.

On the day his relationship with his partner unexpectedly broke down, a friend received news that he had won a place in the finals of a prestigious speaking competition in California. As an introverted expat who had lived in several countries and far flung places, he had been a lonely wanderer before he met his girlfriend. Their relationship had been brief, yet he was filled with anguish that the isolation and loneliness he now felt would become his reality once again. As he began to work on his presentation, he couldn't get past the topic of loneliness and so decided to use the time allotted to him in the competition to showcase his experience of it. He had reached this point in the contest because he had led an interesting life, resulting in many inspirational stories; this wasn't one of them, though, and could well have cost him the prize. My suggestion to him that an emotional frame might be clouding his judgement was dismissed. I didn't know if he was making the wrong choice, but his decision making was lacking. He had decided without acknowledging the influence of his emotions, confirmation and recency bias. His judgement was clouded. I suggested that he may want to outline a second topic as well and test this with speaker colleagues, but he didn't want to.

Once attached to decision targets, integral emotions can have undue influence over our judgement.[14]

Specific emotions influence decision making

If we met up for a drink one evening and ordered the same cocktail from the bar, the chemicals in that cocktail would interact with our unique physiology and influence our mood and bodies in subtly different ways. The second cocktail would amplify that effect. By the third one—well, I'll leave that up to your imagination. Emotions can be thought of as different chemical or hormonal cocktails, each following its own recipe. The latter interact with long-term and working memory to create states of being (or arousal) that affect both *what* we think and *how* we think.

If emotions have hormonal signatures that impact us in consistent ways, it becomes easier to understand the nature of that impact and its influence on our decision making. It is thought that emotions serve a coordination role in our bodies,[15] triggering a set of responses that enable us to respond quickly to problems or opportunities. Specific emotions, like lines of code in a computer program, carry specific 'action tendencies'[16] that signal a universal response to situations, such as anger triggering aggression and disgust triggering avoidance. These are goal-directed responses[17] that influence our judgement and decision making until the situation that caused the emotion is resolved. Not only does an emotion provoke an internal response, but it then goes on to act as a lens through which to see or appraise future events. When we are fearful, risks are amplified, whether this is justified or not. When we are angry, we tend to have increased faith in our ability to influence outcomes and so downplay risks.

I had a coaching client, let's call her Koomala, who was a very senior executive who had moved to a new company after 12 years at her old firm. She moved because she was offered the opportunity to start up a new internal division, with three years of funding to see her through the start-up phase. Four months into building her new team and capability there was a change in management alongside new corporate priorities, and her division was deprioritised. It was then that I started working with her. Needless to say, she was angry. She felt let down by the person who had hired her and the ease with which the commitment made was broken. I quickly learned that this anger had plenty of fuel to sustain itself. Week after week we had conversations in which she illustrated how the business 'no longer supported her' and that she believed she would be made redundant soon. She was able to find kindling to sustain these emotions in almost every interaction she had with colleagues—an offhand comment made in the lifts, not being invited to a social meet-up, secretaries being reshuffled or emails taking longer than they should to be answered.

The influence of anger

In my role as a decision coach I have to focus on the decisions someone is making, not how I feel about them as a person, but it became hard to work with Koomala. I consistently felt like she filtered all information through a persecution complex. I don't give up very easily, though, and began to research how anger impacts our view of the world. What kind of a frame does anger create for us? How did I know that her overriding feeling was anger? Because she said it again and again and my notes of our conversations were filled with the word. In several studies I found answers that changed the way we understood the role of anger and I was able to help her focus on her feelings about various situations rather than on what her response should be.

Imagine cooking up three pots of soup. Each soup has a different broth as its flavour base. Perhaps one is chicken broth, another is beef and a third is vegetable broth. You can go on to add exactly the same vegetables, meat and spices to each soup, but each will have a subtly different flavour, depending on the broth used. Like a broth, emotions taint how we process information and experience events. Each emotion adds a specific flavour and through the work of several research scientists we continuously know more about the subtleties of each.

Koomala's anger was leading her to believe that her troubles were caused by others; it was decreasing her ability to accurately view risks and, perhaps, even take on more risk than she should. Despite this, she remained positive about her own career and her anger drove her to take action. Even though she believed others had caused the problem, she was going to solve it. In her feelings and actions, she recognised the flavour that anger brings to our thinking and behaviour—it reduces sensitivity to risk, increases our beliefs that others have caused our troubles and increases our confidence that we are able to impact the solution. In short, it turns us into fighters[18] in whatever way our personality reflects that. Think of road rage or crimes of passion. Both embody these qualities at the extreme.

The influence of fear

If, instead, Koomala had feared losing her job above all else, her response would have been different in predictable ways. When it comes to risk perception, fear leads to feeling less in control of situational outcomes, and so we perceive more risk in the same situation than those who are angry. This nudges us to play it safe, perhaps even safer than we need to. These emotional impacts influence our ability to accurately judge information, data and even other people.

The influence of sadness

Like fear, sadness increases the tendency for one to perceive events as outside of one's control and so powerless to affect outcomes, even when the reality is quite different. Unlike fear, studies[19] have shown that sadness leads to taking on greater risk in search of greater reward in the hope that will shift us out of feeling down. You might recognise loss aversion in sadness. Remember risk seeking in the face of losses and risk averse in the face of gains?

Emotions not only influence what and how we think, but also *how much* thinking we do. *Depth of processing* reflects how much attention we pay to the various elements of a message when under the influence of an emotion. Do we pay more attention to the messenger (person of high status, expert, a friend or foe, president or nobody special) or the message? Do we get caught in the frame and superficial context or plough through that to think more deeply about the content?

Kahneman's *Thinking Fast and Slow* delineates these two forms of information processing. 'Heuristic' or fast thinking is helpful in many situations when we don't have the time to engage in lengthy meaning making; instead the context of information influences our thinking more than the content. We allow our thinking to be largely influenced by heuristic cues (mental shortcuts), such as the expertise, attractiveness, or likability of the source, stereotypes and the length rather than the quality of the message.[20] This also includes anchoring on certain information, allowing us to jump to conclusions without exerting too much mental processing power. Anger, happiness and disgust all encourage such heuristic processing. As you might have guessed by now, sadness and fear do the opposite, prompting depth of thought about the message and its meaning.

I know, that's quite a lot to take in, so below is a table compiled from numerous studies. For much, much more on this fascinating topic I'd recommend a paper by Lerner et al published in 2015, aptly titled 'Emotions and Decision Making' and freely available online in the *Annual Review of Psychology*.

	Anger	Happiness	Fear	Sadness
Sensitivity to risk	Low	Low	High	High
Risk seeking	High	–	Low	High
Believing the situation to arise because of the action of others	High	Low	High	High
Belief in own ability to influence the outcomes	High	Low	High	High
Depth of processing	Low	Low	High	High

Data compiled from Lerner et al (2015)

Emotions as a source of rich data

Chris Voss is a former lead international kidnapping negotiator for the FBI, and a man trained to keep a poker face, hide all traces of emotion and separate people from the problem. Except that his experiences in the field have taught him that emotions play a key role in successful negotiation outcomes. In writing for

Time Magazine[21] he asks, 'How can you separate people from the problem when their emotions are the problem?' He goes on to explain that, 'Emotions are one of the main things that derail communication. Once people get upset at one another, rational thinking goes out the window. That's why, instead of denying or ignoring emotions, good negotiators identify and influence them.'

Wouldn't you agree that with some understanding of how our emotions influence our thinking, we are far more likely to view things as they are rather than as we want them to be?

It was the ancient Greek philosopher Aristotle who nailed it when he said, 'Anybody can become angry—that is easy, but to be angry with the right person, to the right degree, at the right time, for the right purpose and in the right way—that is not within everybody's power.' Three millennia later, we are beginning to understand that true emotional intelligence doesn't begin with controlling or even using emotions, but unlocking the intelligence that our emotions contain. When it comes to decision making, emotions serve as information filters as well as rich sources of data that aren't particularly difficult to extract.

A colleague was working with a law partnership as they reorganised their commercial structure. The lawyers had been accused of being uncommercial in their approach to new clients and continuously accepting unprofitable revenue streams. The solution, they believed, was to send these lawyers on a 'Finance for Non-financial Professionals' training programme. Was this really the problem, though? Lawyers are pretty smart and definitely understand profit, so why would they intentionally sabotage their own bonuses by choosing quantity over quality? In explaining the situation, my colleague highlighted that this was a rather emotive issue at the firm, leaving several employees hot under the collar. Finance staff felt frustrated and ignored by the lawyers, while the latter felt stressed, overworked and frustrated by having to attend a clearly pointless tick-box training exercise.

My question was: can these feelings be depicted in data? If so, can that data be used to inform the solution? Are the lawyers really ignoring the finance team and their recommendations? Are they doing this because they are truly stressed and overworked or just being recalcitrant?

They set about gathering data to support these emotions by looking for evidence of stress and overwork, such as increased staff turnover, absenteeism, overtime and medical and unpaid leave by the lawyers. Had they taken on more clients recently or received less internal support for them?

In fact, the answer to almost all of these questions was 'Yes'. Their feelings of stress and frustration were reflected in the numbers. Overtime had increased, but so had sick days and unpaid leave. Had they ignored the finance team? No, some had responded to the emails with a request to delay the training or change the format into something less time consuming, like an online programme. Others had highlighted their already ample financial training. But why were they taking on more business at lower margins than they could comfortably handle? Perhaps that was the real issue? On further digging, it seemed that their KPIs were recently changed by HR at HQ, and were now weighted towards new business flows rather than profit.

When we make decisions, we almost always have a feeling about the options at hand. Just a hunch that usually doesn't have much hard data to support it. What do you do with that feeling? How can you use it as a data point in your considerations?

We can't always back up our feelings with data, though. Sometimes we have one of those I-just-don't-think-this-is-a-good-idea moments. It's just a gut feel. Sometimes something doesn't feel right, but you just can't put your finger on it. Gut feel plays a starring role in our decisions, either formally or informally, and my research and that of others bears this out; yet we seldom hear executives stand up and say that one deal or corporate strategy was chosen over others 'on a hunch'.

It's a different story when working with emergency responders, including firefighters, medics and police officers. Their role requires such rapid decision making that gut feel is a valid and necessary decision input. Just like corporate execs, the more experience emergency responders have, the more accurate their gut feelings are. I like to separate gut feelings into the two ways that most of us experience them: as physical sensations in the body and as unconscious insights presented to us without prompting.

The idea that wisdom is a by-product of age is rubbish. Time lived brings the accumulation of decision outcomes, as we experience both good and bad outcomes to choices made over a lifetime. The quality of that experience and the ability to reflect and grow from it is what brings wisdom—and so it's hard to issue a blanket instruction to 'trust your gut as you age'. Of course, the more experienced we become in a particular field, the more accurate and actionable are our mental insights. Insights that are served up at a fraction of the speed of conscious rumination. An executive with many years of experience usually makes adequate decisions much quicker than a novice, as the former has a larger store of decision outcomes to draw on at an unconscious level— even before considering the facts through rational reflection. Seasoned firefighters can understand the nature of a fire (how hot and how deep) by simply looking at the flames because of years of experience, whereas a novice firefighter must still assess each component separately, which takes longer.

Both emergency responders and fighter pilots are continuously exposed to more information than can be processed at a conscious level and so must rely significantly on their intuition to inform their actions. Making sure their intuition is accurate and useful is not something they can leave to the passage of time and accumulation of experiences. Simulation is the gold standard for training instinct, the more realistic the better. Like emotions, instinct and gut feel don't overly discriminate between fact and fiction; so simulated decisions and trial-and-error judgements increase our store of unconscious wisdom and insights, increasingly known as 'somatic markers'.

Antonio Damasio's somatic marker theory[22] proposes a mechanism by which emotional processes can guide (or bias) behaviour, particularly decision making. He and others propose that emotions have associated feelings in the body (*soma* in Greek). These somatic markers guide our thinking towards the most advantageous outcome. This is especially useful when faced with complex or conflicting choices in which our limited cognitive processes become overloaded.

Your somatic markers have evolved from a lifetime of experiences influenced by your own set of preferences, frames and the idiosyncratic circumstances in which you find yourself. The choices you've made within these circumstances (including how to behave, what to say and do, who and what to pay attention to) have resulted in either beneficial or detrimental outcomes. The memories of these outcomes are, according to this theory, stored in memory with associated physical and emotional sensations. For example: a good outcome to a choice may have been associated with a slight increase in heartbeat and flushing of the skin. This multidimensional memory is then stored to be re-experienced later when faced with similar circumstances.

Somatic memories are recalled unconsciously and far quicker than conscious memories, to influence how you view new choices in the hope of guiding you to advantageous solutions. For everyday decision making, a detailed cognitive exploration of all the available options would likely lead to decision paralysis. We don't have the cognitive processing power (nor time) for all of that, and our brains would be overwhelmed. So somatic markers help us out.

We have more time to make good decisions than firefighters or fighter pilots, but we have access to a similar store of information from our own past experiences to call upon. These multidimensional memories will influence how we evaluate our choices and could even simplify the decision process. You've experienced these feelings before through gut feel, hunches or preferences.

Because we don't know exactly from which experiences these somatic markers developed, I prefer to use them as data points in understanding my choices. If something doesn't feel right, or I just prefer one option over another, I use this as a basis for exploring why. An excellent place to start is by asking: What is it about the option, or about me, that makes me feel this way? This will generate further information that could be either subjectively or objectively substantiated. Either way, it will help you explore shadows in your thinking where information might be incomplete or the risks poorly defined.

As you can see, when it comes to decision making, emotions and conscious deliberation are two sides of the same coin, sharing an interwoven and symbiotic relationship. After all, soup without broth would just be boiled vegetables and pale bits of meat.

14

GENDER DIFFERENCES IN DECISION MAKING

Everything you need to know about how men and women process information differently but were too afraid to ask. What this means for risk perception and team decision making.

It's no secret nor is it discriminatory to point out that men have bigger brains—it's simply biology. On the other hand, female brains are more densely packed, giving men and women more or less the same processing power. Physical differences aside, the real differences in our decision making comes from differences in how the sexes encode memories, sense emotions, recognise faces, take risks and solve certain problems.[23] These gender discrete cognitive functions affect what we think and how we think, just like emotions and mental biases. Is it important to be aware of them? We've done pretty well just muddling along in male or female bodies for 200,000 years, yet being able to capitalise on the different perspectives that this crude form of cognitive diversity yields is an easy win all round. Let me explain.

Information flows differently through male and female brains. We are literally wired differently through what are called cognitive connectome circuits. Like a Google map of all the roads on earth, connectome circuits show us where we go and can travel—but not if we travel by bus or moped or take an Uber, or what the purpose of the journey is. Our understanding of these circuits is in its infancy, yet Madhura Ingalhalikar, Alex Smith and their collaborators[24] were able to map the structural connectome of a large cohort of male and female brains (949 healthy living brains, to be exact). Even though we already suspect that men are generally better with directions and fitting everything into the car for that long road trip because they have superior spatial abilities and that women remember details better than men, these researchers were able to add some science to our hunches. Including why women have better social cognition abilities but men can boast better motor skills.

They found that female brains are optimised for interhemispheric communication. This ability for signals to flow easily across the left and right lobe facilitates communication between the brain's analytical and intuitive processing modes. In contrast, male brains were found to be optimised for intrahemispheric communication—back-to-front and front-to-back, rather than communicating across the hemispheres. This facilitates connectivity between perception and coordinated action. Given this, it's easy to see where the old adage of men as hunters vs women as gatherers comes from.

What does any of this mean for decision making? Think about the areas of the brain that women instinctively recruit in making decisions: intuition and analytical circuits. This seamless connection between thinking and feeling influences what we pay attention to and improves women's ability to integrate emotional and social aspects into analysis and decision making. Perhaps the impact of a woman's choices is naturally explored in a wider context outside of the immediate goals; perhaps nature nudges us to think more about how it will affect others? Here intuition might play a starring role or eclipse the use of reason in a pinch.

For men, receiving information and acting on it is the default wiring. This would make men quicker and more decisive decision makers. Information gathered might be task rather than people focused. Thinking might be action orientated and risks narrowly evaluated. I use *may*, *might* and *could* rather than more definitive verbs, because I am describing average behaviour—and there will be a wide range of behaviours around the average.

Think about your experiences in working with teams made up of both men and women. Did the men seem more decisive, while the women were more thoughtful or circumspect in decision making? Did the women want more information to consider the wider effects on stakeholders, while the men were happy to course correct along the way?

I've seen this many times, but mostly because I am looking for it. I'm looking for ways to ensure that teams recognise cognitive diversity and make the most of the different angles of attack that different genders and nationalities bring to a problem domain.

In addition, research in affective neuroscience[25] illustrates that women are better at recognising emotions and express themselves more easily, and have enhanced information recall compared to males.[26] On the other hand, men show greater responses to threatening cues (dominant, violent or aggressive) and are more inclined to take risks than women.[27] Working together in decision-making scenarios, it's not difficult to see how a woman's recall of facts and ability to read emotions of stakeholders can complement a man's ability to more accurately assess and take risks (and read a map).

I'm often asked if these gender differences can be pinned on nature or nurture. What I've learned is that bifurcation of male and female brains becomes apparent at a young age, demonstrating wide differences during adolescence and adulthood.[28] So, at this stage, the answer is that it could be either; but without a doubt, some research scientist somewhere is looking into it. Watch this space.

15

STRESS IS AN
EMOTION, TOO

Exploring stress as an emotion and its effects on your risk processing—as well as your ability to deal with idiots.

What about the emotion of stress? Getting to know oneself under stress is an ever more important part of making the best decisions we can. When stressed, our senses are heightened from short bursts of stress hormones, maintenance functions in our body shut down and blood redirects to our limbs, priming them for action. Short bursts of stress were all our ancestors needed to escape imminent danger. When the threat had passed, homeostasis and balance could return to their body and mind. Life in the concrete jungle is a little more complicated, though, as our natural state of balance includes the daily hum of low-level stress—stress from traffic, bills, long working hours, bad bosses, information overload, or the reality of not being thin, healthy, wealthy, young or good looking, etc. enough. Under these conditions it doesn't take much to tip us out of our natural balance. How do you know when you are particularly stressed? Think about it right now. Then think about the impact of stress on your decision making. Do you know what it is?

I know I am stressed when:

1 _____

2 _____

3 _____

Stress affects my decision making in the following ways:

1 _____

2 _____

3 _____

Do you believe that stress affects your health?

Yes / No

Stress shortens my temper, reduces my appetite, frustrates my sleep and greatly decreases my tolerance of idiots—and there seem to be many more idiots in my life when I'm stressed. Sound familiar? But that's not the worst part. In case you forgot, let me remind you that chronic stress can lead to high blood pressure, a dampening of the immune system and increased susceptibility to common infections. It contributes to asthma, digestive disorders and cancer—and let's not forget that it ages you more quickly. Pretty scary stuff, isn't it? Fortunately, we are no longer powerless over the ravages of this insidious emotion. The new science of stress is shattering our traditional ideas about it and giving us new strategies in our pushback against it—and I'm not talking about triple wheatgrass shots here.

We've long believed that stress is the leading cause of heart disease, which is the leading cause of premature death in developed economies. A flood of new research shows us that simply believing this statement to be true raises the odds of your early demise by 43 per cent.

Huh?

People who don't believe that stress is harmful to their health have been found to experience less adverse long-term health effects from elevated levels of stress. In a large, eight-year study, people who experienced a lot of stress but did not view stress as harmful had the lowest risk of dying prematurely of everyone studied, even compared to those who had relatively little stress. A team of researchers at Harvard and the University of California wondered how this was even possible and explored the physical consequences of changing our minds about the effects of stress on our body. What they found was fascinating.[29]

For the average Joe on an average day, stress will ramp up his heart rate while causing his blood vessels to constrict; but in this study, participants were encouraged to view their stress response as helpful and preparing them for the challenges ahead. With this new mindset, their heart rate still went up when they experienced stress, but their blood vessels stayed relaxed. The resulting cardiovascular profile looked a lot like that of someone experiencing a moment of courage or bravery in the face of adversity. This is much less damaging than the typical stress response and can explain why a lifetime of believing that stress is helpful leads to a healthier body.

It would seem that viewing stress as a sign that our body is energised and preparing us for action can counteract the physical toll it would otherwise take—almost like a mental vaccine against stress. If you believe that your stress response will help your performance, then you can expect to be less anxious and more confident, and don't forget, also less likely to succumb to stress related ailments.

> "Over a lifetime of stressful experiences,
> this one biological change could be the
> difference between a stress-induced heart attack
> at age 50 and living well into your 90s.
> This is really what the new science of stress reveals,
> that how you think about stress matters."
>
> —Kelly McGonigal, health psychologist,
> lecturer and researcher at Stanford University

But what about the effect of stress on decision making? The latter can be viewed through its effects on our ability to accurately perceive and take on risk. In the previous chapter we saw some biological differences between men and women that would affect our information processing and decision making in predictable ways. Using stress as a filter through which we view information, researchers have found that male and female brains respond differently in this domain as well.

Quite simply, when decisions are presented under conditions of uncertainty, multiple studies have reported reduced risk taking in stressed females but increased risk taking in stressed males. This holds true, even when taking the risk is clearly advantageous or disadvantageous,[30] concluding that acute stress increases risk taking in men while decreasing it in women.

Given what we learned about how information flows through the brain, this isn't too hard to believe. To recap: information travels through social and analytical areas more readily in female brains, whereas men are wired for acting on incoming information more directly. I seldom get pushback from audiences on this point, except once when I was working with a large group of senior leaders from a FTSE 100 manufacturing company. I was surprised by the number of women in senior positions who clearly held real influence over outcomes. This group thought gender differences in risk taking was nonsense, because the women at this company were as capable as the men in taking on risks.

It soon became clear that, in this company, as women rose to the top, risk-taking behaviour became an internal norm within leadership. They were selected for it without knowing it, and those women would continue to promote women with risk-taking personas reflecting their own.

There is further breaking news on the impact of stress on our decision making. Stress hormones are an essential partner in life, supporting us in rising to daily challenges and overcoming obstacles en route to achieving our goals. However, too much stress with no clear end in sight results in an acute condition where we are bombarded daily with high levels of stress hormones. Such acute stress changes the impact of stress hormones and affects our ability to make appropriate decisions. Some poor lab rats were exposed to chronic, unpredictable stress over time and their behaviour changed quite dramatically.[31] Specifically, they became resistant to change as their behaviour rapidly shifted from goal-directed (i.e. searching broadly to find food) to habitual strategies (methods that had been successful in the past) that proved impervious to bigger and better rewards.

Now imagine if this behaviour is replicated in humans—whole organisations of chronically stressed individuals, unmotivated by ever-larger rewards and less able to adapt behaviour to meet challenges and goals. Employees simply going through the motions every day, resisting change and avoiding goals and behaviours that stretch them out of their comfort zone. Better cafeterias, comfortable bean-bags and ping-pong tables are not always the answer to lowering employees' stress.

16

THE PAIN AND POWER
OF ALTERNATIVE OPINIONS

Alternative opinions as data points. What are the key
differences between novice and expert decision makers?
Why criticism is hard to receive and even harder to give.
Not everyone qualifies to criticise you,
so choose your critics wisely.

We met Ray Dalio in chapter 7 when discussing process-orientated
decision making. A more formal introduction is required at this
point, so please meet Ray Dalio again, a 69-year-old who still draws
crowds any ageing rockstar would be envious of. He doesn't sing,
dance or try to motivate you; instead, he grabs your attention by
telling you how it is. Or at least the way it was for him and how he
built his company and fortune through non-mainstream decision-
making methods, such as radical transparency and unflinching
candour. He is the founder of investment firm Bridgewater
Associates, one of the world's largest hedge funds, and the world's
58th wealthiest person on Bloomberg's June 2019 list.

Dalio has many, many principles that he believes have contributed
to his success and he explores them all in his book *Principles*. But
he has one special talent that I see very rarely in decision makers

and which I believe a large part of his success derives from: quite simply, his ability to seek out and accept criticism.

In his own words:[32]

> "In trading markets there's a high probability that you are going to be wrong no matter how confident you are.
> So I learned early on to look for people that had alternative points of view ... people who disagreed with me—because I wanted to see their thinking.
> This improved my chances of being right.
> I learned a lot from the perspectives of others.
> I think we're all blind in many different ways."
>
> —Ray Dalio in an interview at the
> *New York Times* DealBook Conference

Sounds pretty easy, right? Find someone who knows a bit about the subject you're thinking about, book some of their precious time, explain your ideas and ask them where they would disagree with you. Listen carefully and incorporate their ideas as data points to be explored in your thinking. Maybe this isn't such a tall order for you? Perhaps you have a network of trusted advisors? Most folks I work with use their friends and friendly colleagues to bounce ideas off. However, it's simply not as easy to get this step right as it sounds. For starters, working with friends is not biologically optimal. Here's why:

1. Giving criticism to a friend or friendly colleague goes against our nature.
Friends tend to be like-minded people who enjoy each other's company and value each other's friendship at a neurological level.

Friendship activates our brain's reward circuits, creating a feeling we enjoy and wouldn't easily sabotage. On the other hand, asking a friend to disagree with you can increase activity in their amygdala (a part of the inner brain associated with triggering your brain's fear system). This can produce such discomfort that your friends would rather avoid giving criticism or downplay it. It's almost unfair to put them through such emotional stress.

2. Receiving criticism from anyone goes against our nature.
But receiving criticism isn't a picnic for you, either. When you and someone you consider a friend or trusted advisor are in agreement, two particular brain areas respond: the nucleus accumbens, part of the same reward circuitry of the brain; and the orbitofrontal cortex, the portion of the brain involved in the cognitive aspects of decision-making.[33] On the other hand, when that bond is interrupted through disagreement, researchers note that such exclusion causes emotional pain that follows the same neural circuitry as physical pain.

As you can see, we just aren't wired to give or receive criticism painlessly, no matter how constructive or well-intentioned it is. Much of the research in this area is linked to conformity and the need to be accepted by your peers. Taken to an *extreme*, this is foundational to the destructive nature of herd behaviour and groupthink—behaviours that dramatically impact sound decision making.

Dalio saw alternative options as a way of helping him be less wrong. Being less wrong than other investors leaves you in the money. For him, it wasn't criticism, it was vital information. His ability to reframe disagreement made all the difference to him.

Here are some ideas to ease the cognitive impact of receiving (and giving) criticism, disagreement or challenge to your ideas.

Reframe from criticism to data
Think of reaching conclusions as building a puzzle as best you can. You hold some of the pieces, some pieces may be completely missing, while others hold a few key pieces. Soliciting puzzle pieces, in the form of different perspectives and alternative opinions, will help to clarify the picture on your puzzle. Of course, you could recreate the pieces yourself, but this limits you to your own perspective, takes even longer and concentrates your mental biases.

Ask early in the process
Good decision making takes a bit of time, effort and mental investment. Once you've reached your conclusions, you have developed a relationship with them, they are the product of your skill and acumen. Once you own a conclusion, it is fully subject to both the sunk cost effect and loss aversion (where the fear of losing something you own outweighs the joy of receiving something of equal value). Try not to present your conclusions to others for scrutiny, but rather invite others to comment on the thinking that will generate your conclusions early on in the process.

Choose your critics
Not everyone is qualified to criticise you. I made my 13-year-old son his first coffee the other day as he struggled with concentration levels during exams. He hated it. I wasn't surprised, and certainly

didn't take it personally; he was hardly in a position to judge my barista skills.

Given social media today, there are so many strangers out there who will criticise you without experiencing any neurological malaise. In the public work that I do, I constantly remind myself that not every critic has earned the right to criticise me. At Bridgewater Associates (Dalio's firm) they utilise proprietary software that applies a 'believability weighting' to opinions and analysis. Those that have the most experience and best track record on the topic under discussion will have their opinion upweighted in the final analysis. They have earned the right to give an opinion.

Given that challenge and criticism have a far greater influence on our thinking[34] (and well-being) than positive comments, critics have a far greater impact on your decisions and life in general than your friends do. And so, might I suggest that choosing our critics wisely is even more important than choosing our friends and allies?

Don't seek approval, seek information

When I was pregnant with my son (the now 13-year-old who is not quite yet a coffee lover), I was vegetarian and had been for over a decade. It worried me that, perhaps, my dietary preferences would impact the health of my baby. I approached the subject with my gynaecologist, who offered to refer me to a nutritionist, but not before saying, 'If I were you, I would just start eating meat again, it's what a baby needs. It's unfair to impose your will on an unborn child.' Oh my, did I feel like a lousy, selfish mother already and I had only been pregnant for a couple of weeks. Two weeks later, after a full blood and dietary analysis, the dietician showed me how to give my baby all the nutrients he needed from a plant and dairy-based diet. Despite having a beautiful, healthy baby and now teen, that taunt from the gynaecologist has stayed with me all these years. It no longer stings, it just reminds me that her area of expertise was pregnancy, not nutrition. The nutritionist was able to give me facts and allow me to make up my own mind.

Opinions, like feelings, are generally subjective; getting to the facts that they are based on is far more useful in our decision making.

PART 4

17

YOUR ENHANCED
DECISION-MAKING PROCESS

Your new decision-making process. A quick summary
and DECIDE Decision Making Cards for teams.

When we started working on your decision making, I asked
you to write down your process. Can you now write down your
augmented process, including some or all of the tools that we
explored in this book?

As a reminder, they are:

- a process rather than outcome orientation
- clear decision rights—checking who is the decision maker
- a meta-decision (including ensuring that the correct problem is
  being solved)
- checking that the problem is being correctly framed
- being aware of the mental biases you/your team are most
  prone to
- exploring assumptions and risk
- understanding the role of unconscious processes on risk perception
- gathering challenging opinions

Given the above, what would your ideal decision-making process look like now?

Chapter summaries

Part 1
Chapter 1 introduces the question of: What is a good decision and how should it be evaluated?

Chapter 2 explores a crowdsourced explanation of the most common answers to the above question. These answers come from a decade of asking this question in both formal research and in both corporate and academic engagements.

Conventional wisdom tells us that a good decision:

- achieves its objectives
- logically weighs all the options at hand
- avoids thinking clouded by emotions
- aligns to the organisation's or individual's goals and values
- avoids regret

Chapters 3 to 5 challenge each one of these points in turn to determine their role in good decision making. We conclude through theory and practice that not all of these are relevant to making a good decision as follows:

- *Achieves its objectives*—No, this isn't a necessary criterion for a good decision.

- *Logically considers or weighs all the options at hand*—One cannot possibly weigh all the options available, and traditional definitions of logic or rationality are at odds with how we process information and make decisions. An updated functional definition of rationality is suggested.

- *Avoids thinking clouded by emotions*—Given the role that emotions play in decision making, namely that they influence what we pay attention to, how we gather and process information and how we evaluate risks, it is biologically impossible to make decisions without the influence of emotions.

- *Aligns to the organisation's or individual's goals and values*—This is only a valid decision-making criterion if those goals and values are periodically examined to avoid the danger that they become unhelpful and outdated lenses through which we view data and information.

- *Avoids regret*—After testing all the candidate theories against academic and practitioner research, we are able to conclude that a good decision is one that the decision maker doesn't regret. Freedom from regret hails from making the best possible decision one can with the available resources (physical, mental and time) and a realistic understanding of one's limitations. The best way to be sure of achieving this is using a decision-making process that guides one's thinking, especially when under pressure. How to do this is the subject of Part 2.

Part 2

In Part 2, chapters 6 to 10 introduce a best practice decision-making process and begin to explore and explain the steps that it consists of alongside examples of decision-making processes, protocols and debiasing strategies used in organisations today.

If you are a professional of any kind, from a tech engineer to an HR executive, a CEO or a dentist, you are continuously weighing up options and deciding on the best trade, payoff, treatment or even the best thing to say in a presentation or to a customer. You are a professional decision maker and your success depends largely on the quality of your decisions. You've already learned that a good quality decision isn't always the one with the best outcome. What's far more important than hitting the bull's eye every time is to foster a good decision process that ultimately results in incrementally better decisions and hence gains from those decisions over time.

Decision processes are highly personal, but a best practice decision-making process should include some of these success factors:

A process rather than outcome orientation

Good decisions are never random inspirations hastened by a moment of genius or lucidity. A process is used (consciously or subconsciously) by anyone who makes consistently good decisions, because no one is consistently lucky. Chapter 6 asks you to write down your process so that you can reflect on and refine your approach to problem solving as we go through the tools introduced.

Clear decision rights

Once you know you are solving the correct problem, it's a good idea to clarify who the decision maker is for the various aspects of the problem domain and ensure that they have both the authority and resources to exercise their right to take decisions. What choices are under your control and what choices need decisions from others as inputs? Each decision maker should understand their risk budget, know how much risk they are at liberty to take, and know what resources can be allocated to their decision-making efforts.

A meta-decision

A meta-decision is the simple act of deciding *how* you will decide before you jump in and make a decision. It begins by checking that you are, in fact, solving the *right* problem, then asks you to decide how you will solve the problem—using which tools, data and resources. It sounds like a mini project plan because it is. The meta-decision forms the very first step in a good decision process because it anticipates challenges, ensures that you are using the best possible tools, that your team members are all on the same page and actually speeds up the decision process.

Checking that the problem is being correctly framed

It was Socrates who first proposed that all information occurs within points of view and frames of reference and that all reasoning proceeds from some goal or objective. The poor man was executed for his outrageous thinking. Today this reasoning separates good decision makers from the rest. Without fail, every piece of information that is presented to you is done so through someone else's frame of reference, and hence has been structured in a way that serves their ends. Always ask yourself what motivation the journalist, stockbroker, surgeon, CEO, or any other person has when transmitting information. If you have sourced data yourself, also beware—that data is filtered through your own mental frames.

Being aware of the mental biases you/your team are most prone to
Outside of finance and advertising, where executives can use psychology to profit from their customers' biases, in strategic decision making, executives must understand and counteract their own biases and those of their colleagues. There aren't many examples of this being done successfully, with recent research on the tongue-twisting *bias blind spot bias* reminding us that most of us tend to perceive ourselves as less susceptible to biases than others. Training to reduce explicit biases has also not proven to be as successful as expected in eliminating them. Again, this is because we find it hard to believe that we are explicitly biased against others. This chapter focuses instead on implicit bias or mental shortcuts that affect how we process information and perceive risk, such as loss aversion, confirmation bias, anchoring and overconfidence.

Part 2 ends with an exploration of debiasing strategies. Knowing that a bias possibly affects your thinking doesn't guarantee that it won't. Debiasing strategies are decision protocols or frameworks that counteract the most prevalent biases in your thinking or that of your team. See examples of such strategies used in organisations at the end of chapter 10.

Part 3
The remainder of a best practice decision-making process is explored in Part 3 and includes:

- exploring assumptions and risk
- understanding the role of unconscious processes on risk perception
- gathering challenging opinions

In chapter 11, we see how assumptions play a starring role in our decisions. These are sometimes born from past data or the status quo and sometimes from what we don't know or beliefs we haven't updated. They are stories we create and repeat so often that they sometimes graduate to 'facts' in our thinking. Part 3 explores strategies to root out assumptions, such as asking the following:

- What do I/we know, but can't prove?
- What do we accept without challenging? What is the status quo?
- What do we not know?

Decision making usually begins with an inside-out process of understanding the decision, our options and their impact on us. Questioning base rates and assumptions allows us to start with an outside-in view, which has a wider frame and a less personal lens.

Exploring decision variables and assumptions in a *States of the World Matrix* is a useful tool to qualify uncertainty, as explored in chapter 11:

	State of the world 1	State of the world 2
Action 1	Outcome 1	Outcome 2
Action 2	Outcome 3	Outcome 4

In chapter 12, we take a well-earned break from exploring our own decision making and focus on that of the FBI instead, with a true case study exploring the wrongful arrest and imprisonment of Brandon Mayfield. Once you've seen their decision-making mistakes in action, you can make recommendations to help them get better and see if your recommendations align to those suggested by an independent commission investigating the case.

Chapters 13 to 15 turn to the mega influence of unconscious processes on our risk perception. It is thought that emotions serve a coordination role in our bodies, triggering a set of responses that enable us to react quickly to problems or opportunities. Specific emotions, like lines of code in a computer program, carry specific 'action tendencies' that signal a universal response to situations that influence our judgement and decision making. Not only does an emotion provoke an internal response, but it also acts as a lens through which to see or appraise future events. This is why it is essential to check in with how you feel about the decision and its components. Stress is an emotion, too. We explore the impact of stress on risk perception along gender-specific lines and breaking research on the ability of stress hormones to bias decision making away from goal orientated to habitual behaviours.

Chapter 16 reminds you that testing your thinking before making a decision is an important part of good decision making; yet we aren't wired to give or receive criticism painlessly, no matter how constructive or well intentioned it is. Reframing from criticism to data, asking for challenging opinions early on in your thinking process, and choosing your critics even more wisely than your friends all help to manage and maximise the impact of receiving criticism.

Finally, chapter 17 concludes with a summary of a best practice decision-making process and a spot for you to write down how you will use this to augment the decision-making process that you started out with.

DECIDE™
DECISION MAKING CARDS

Today, more often than not, decisions are made in teams. A great way to encourage structure in team decision making is to use DECIDE™ Decision Making Cards. Not every aspect of a decision-making process is needed when making a decision and these cards allow teams to pick the ones that are most applicable for them. Cards can then be distributed to support various team members in leading each part of the conversation. Here is an example of the DECIDE cards, but you can draw up your own cards like you would cue cards for a speech, or flash cards for an exam (remember those?).

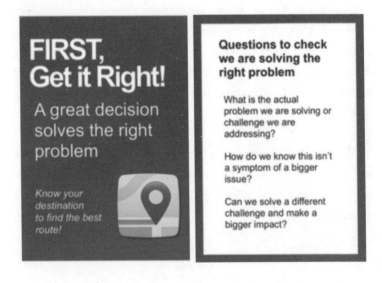

DECIDE
How to Decide

A **Metadecision** checks on
- Framing
- Resources available
- Assumptions made
- Method to be used
- Past experience

Universal settings for better decision outcomes

PAST EXPERIENCE

Is there another industry/company/department/person who has faced a similar challenge?

How did they solve it? What worked for them? What didn't?

What can we learn from their experience?

MANAGE
Blindspots & Emotions

Emotions and mental blindspots affect what and how I think

90 Seconds is all it takes for an emotion to pass!

Managing the impact of emotions and blindspots

How am I feeling right now?

How could this affect my perception of risk?

Is there information I am anchoring on?

Am I gathering information that agrees with my thinking as well as data that challenges it?

OVER TO YOU

Introducing tools to improve decision making has had a profound impact in many parts of the world, from NGOs working in the Middle East, to rewriting history textbooks to make them more accurate and less biased, to reducing plastic waste and enhancing ethical decision making in Big Pharma. It's helped investment teams engage in more robust debate, improved returns on investment for various firms and it's helped me weather many personal and professional challenges over my career and life. I can only hope that, whatever form your new, enhanced decision-making process takes, it helps you live a little more bravely and with less regret, knowing that you've made the best possible choices.

Remember that the decisions you made in the past, and how you've chosen to react to what life throws at you, have resulted in your current reality; the decisions you make from today, will create your future.

Choose wisely.

AUTHOR'S NOTE

This book was written as part of my PhD thesis. The aims of the latter were:

1. to contribute to the theory of decision making through a case study that examines the creation of a behavioural decision-making process, and
2. to improve decision making in practice.

I realised that an academic thesis was not, in any way, going to improve decision making in practice and so extracted the best ideas from a very long academic submission and wrote them up in *DECIDE*. Informed readers will recognise this book in the thesis and vice versa.

REFERENCES

Part 1

[1] *Nudge* is narrowly focused and wouldn't qualify as a general treatment of decision making and so has limited applications, despite large helpings of food for thought.

[2] My current research covers behavioural decision systems/frameworks that can be used across organisations to improve the quality and ethics of organisation-wide decision making. Yes, there will be another book to cover my research, and I promise it won't be overly academic.

[3] From Pascal's *Pensées* Part III—'The Necessity of the Wager' (Trotter translation), available at Classical Library (Wager found at #233). Blaise Pascal was a 17th-century French philosopher, mathematician and physicist.

[4] https://en.oxforddictionaries.com/definition/rationality retrieved on 28/11/2018: 'logic'

[5] https://en.oxforddictionaries.com/definition/rationality retrieved on 28/11/2018: 'rationality'

[6] Simply multiply the payoff by the probability of it occurring, e.g., 400 x 0.2 = 80

[7] Bernoulli could not have been a Nobel laureate in the 1800s but his work, nevertheless, inspired several award-winning theories.

[8] The original currency quoted was a European trading currency consisting of gold, silver and other metallic coins called ducats.

[9] Probability: the extent to which an event is likely to occur, measured by the ratio of the favourable cases to the whole number of cases possible; see https://en.oxforddictionaries.com/definition/probability.

[10] The development of a rational decision theory was first posited by mathematician John van Neumann and economist Oskar Morgenstern in 1953, who offered a mathematical theory of decision making underpinned by Bernoulli's principle of maximising expected utility or rewards that may differ from the monetary value of a gamble. They explored the conditions under which the expected utility hypothesis would be valid.

[11] Quoted from Lewis, M. (2016) *The undoing project: a friendship that changed the world*. UK: Allen Lane. A favourite book of mine.

[12] In 1956 Herbert Simon suggested that since we suffer from limited computational facilities and are almost always subjected to limited information, we can be expected to employ an 'approximate' form of rationality that he called bounded rationality or satisficing. It describes how we strive for choices that return a satisfactory outcome and stop searching when we believe we have found this, rather than continuing until we have reached the optimum outcome, if such a thing exists.

[13] Conlisk, J. (1996) 'Why bounded rationality?', *Journal of Economic Literature* 34(2): pp. 669–700, p. 692

[14] Gigerenzer, Gerd (2008) *Rationality for Mortals: How People Cope with Uncertainty*. Cary: Oxford University Press, Incorporated.

[15] Hastie, R. and Dawes, R.M. (2010) 'Rational choice in an uncertain world: the psychology of judgment and decision making'. 2nd edn. Los Angeles: SAGE.

[16 & 17] Joseph le Doux (2015) 'Feelings: What Are They and How Does the Brain Make Them?' *Daedalus*. MIT Press, 144(1): pp. 96–111

[18 & 19] Keltner D., Lerner J.S. (2010) 'Emotion', *The Handbook of Social Psychology*; ed. D.T. Gilbert, S.T. Fiske, G. Lindzey: pp. 317–52. New York, NY: Wiley; and Lerner, J.S., Li, Y., Valdesolo, P. and Kassam, K.S. (2015) 'Emotion and Decision Making', *Annual Review of Psychology*, 66(1): pp. 799–823

[20] Lerner, J.S. and Keltner, D. (2001) 'Fear, Anger, and Risk', *Journal of Personality and Social Psychology*, 81(1): pp. 146–159

[21] If that relationship ended on a very negative note then you might not remember the full glow and glory of its positive beginning as these memories would have been tainted.

[22] Galentino, A., Bonini, N. and Savadori, L. (2017) 'Positive Arousal Increases Individuals' Preferences for Risk', *Frontiers in Psychology* 8: pp. 21–42

[23] Lerner, J.S. and Keltner, D. (2001) 'Fear, Anger, and Risk', *Journal of Personality and Social Psychology*. American Psychological Association, 81(1): pp. 146–159

[24] *The Hour Between Dog and Wolf: Risk-Taking, Gut Feelings and the Biology of Boom and Bust*, by John Coates (2012) is a fascinating exposé of the impact of emotions on decision making during a financial crisis and well worth a read.

[25] Reavis, Rebecca and Overman, William H. (2001) 'Adult Sex Differences on a Decision-Making Task Previously Shown to Depend on the Orbital Prefrontal Cortex', *Behavioral Neuroscience*. American Psychological Association, 115(1): pp. 196–206; and van Honk, Jack *et al.* (2004) 'Testosterone shifts the balance between sensitivity for punishment and reward in healthy young women', *Psychoneuroendocrinology*. Elsevier Ltd, 29(7): pp. 937–943

[26] Britain's exit from the European Union was supported by 51.9 per cent to 48.1 per cent of the UK population that voted in a referendum in 2016.

[27] Gilead, M., Sela, M. and Maril, A. (2018) 'That's My Truth: Evidence for Involuntary Opinion Confirmation', *Social Psychological and Personality Science*

[28] Gilead, M., Sela, M. and Maril, A. (2018) 'That's My Truth: Evidence for Involuntary Opinion Confirmation', *Social Psychological and Personality Science*, p. 8

[29] This would include the behaviours that underline that ethical conduct, for example: being trustworthy, which could be defined as saying what you mean, meaning what you say and doing what you commit to. Behavioural definitions have a greater granularity than the semantic definitions.

[30] See Tremaine du Preez, Middlesex University (2020)

[31] Enron, Annual Report (2000) p. 29

Part 2

1 Lovallo, D. & Sibony, O. (2010) 'The Case For Behavioural Strategy', McKinsey & Company

2 Gideon, K. and Bruine de Bruin, W. (2003) 'On the Assessment of Decision Quality: Considerations Regarding Utility, Conflict and Accountability', in Hardman, D. and Macchi, l. (eds.) Chichester, UK: John Wiley & Sons, Ltd: pp. 347–363

3 Available at Investor's Archive: https://www.youtube.com/watch?v=5x3TdtjLibM

4 https://www.investopedia.com/investing/warren-buffetts-investing-style-reviewed/

5 'The Manufacturing Manager's Skills' (1966) by William H. Markle (Vice President, Stainless Processing Company, Chicago, Illinois), Quote pg. 18. Published by American Management Association, Inc., New York.

6 Those promoting the UK's withdrawal from the European Union.

7 Available at https://www.ons.gov.uk/

8 This example was inspired by an article in *The Economist* published on 4th May 2019 titled 'Netflix and pills. The antibiotic industry is broken.'

9 Russo, J.E. and Schoemaker, P.J.H. (2001) *Winning Decisions: Getting It Right the First Time*. 1st edn. New York: Bantam Doubleday Dell Publishing Group.

10 Arkowitz, H. and Lilienfeld, S.O. (2009, January 8) 'Why science tells us not to rely on eyewitness accounts'. *Scientific American*.

11 The somatic marker hypothesis (SMH) proposes a mechanism by which emotional processes can guide (or bias) behaviour, particularly decision making. Also see: Damasio, Antonio R. (2008 [1994]) *Descartes' Error: Emotion, Reason and the Human Brain*. Random House.

12 Atewologun, D. et al. (2018) 'Unconscious bias training: An assessment of the evidence for effectiveness', Equality and Human Rights Commission, Research Report 113. Available at www.equalityhumanrights.com

13 'How to Make Better Decisions: Prospect Theory on BBC Horizon's Show', 2012

14 Lewis, M. (2016) *The Undoing Project: a Friendship That Changed the World*. UK: Allen Lane, p. 269

15 Gignac, G.E. and Zajenkowski, M. (2019) 'People tend to overestimate their romantic partner's intelligence even more than their own', *Intelligence*, 73: pp. 41–51

16 Interview with Daniel Kahneman: 'What would I eliminate if I had a magic wand? Overconfidence' by David Shariatmadari, published in *The Guardian* newspaper online on 18/07/2015, retrieved on 26/03/2019

17 INET interview with David Tuckett: 'How Stories about Economic Fundamentals Drive Financial Markets', *New Economic Thinking*, published on 13/01/12 retrieved on 26/03/2019

18 Pronin, Emily, Lin, Daniel Y. and Ross, Lee (2002) 'The Bias Blind Spot: Perceptions of Bias in Self Versus Others', *Personality and Social Psychology Bulletin*. Thousand Oaks, CA: Sage Publications, 28(3): pp. 369–381; and Chandrashekar, Prasad et al, (2019) 'Agency and self-other asymmetries in perceived bias and shortcomings: Replications of the Bias Blind Spot and extensions linking to free will beliefs'. Published on Researchgate—10.13140/RG.2.2.19878.16961

Part 3

1 & 2 Shariatmadari, D. (18/07/2015) Interview with Daniel Kahneman: 'What would I eliminate if I had a magic wand? Overconfidence', Guardian online

3 Nicas, J., Kitroeff, N., Gelles, D. and Glanz, J. (01/06/2019) 'Boeing Built Deadly Assumptions Into 737 Max, Blind to a Late Design Change', *The New York Times*

4 Gigerenzer, G., Hoffrage, U. (1995) 'How to improve bayesian reasoning without instructions: frequency formats'. *Psychology Review* 102, pp. 684–704

5–8 Office of the Inspector General Oversight and Review Division (2006) 'A review of the FBI's handling of the Brandon Mayfield case', Unclassified Executive Summary, US Department of Justice. https://oig.justice.gov/special/s0601/final.pdf. pp. 8–12

9 Darwin, C. ([1972]1965) *The Expression of the Emotions in Man and Animals*. Chicago, University of Chicago Press.

[10 & 11] Phelps, E., Lempert, K. and Sokol-Hessner, P. (2014) *Emotion and Decision Making: Multiple Modulatory Neural Circuits*, Palo Alto: Annual Reviews, Inc.; and Le Doux, J. (2015) 'Feelings: What Are They & How Does the Brain Make Them?' *Daedalus* 144(1): pp. 96–111

[12 & 14] Lerner, J.S., Li, Y., Valdesolo, P. and Kassam, K.S. (2015) 'Emotion and Decision Making', *Annual Review of Psychology*, 66(1): pp. 799–823

[13] Damasio, Antonio R. (2008 [1994]). *Descartes' Error: Emotion, Reason and the Human Brain*. Random House.

[15] Frijda, N.H. (1993) 'The Place of Appraisal in Emotion', *Cognition and Emotion* 7 pp. 357–387

[16] Frijda, N.H. (1986). *The Emotions*. Cambridge: Cambridge University Press

[17] See Lerner & Keltner (2000), Lerner & Keltner (2001)

[18] Frijda, N.H. (1988). 'The laws of emotion'. *American Psychologist*, 43: pp. 349–358

[19] Lerner, J.S., Li, Y., Valdesolo, P. and Kassam, K.S. (2015) 'Emotion and Decision Making', *Annual Review of Psychology*, 66(1): pp. 799–823

[20] Bless H., Schwarz N., Clore G.L., Golisano V., Rabe C., Wolk M. (1996). 'Mood and the use of scripts: Does a happy mood really lead to mindlessness?' *Journal of Personality and Social Psychology* 71: pp. 665–79; and Bodenhausen G.V., Kramer G.P., Süsser K. (1994). 'Happiness and stereotypic thinking in social judgment', *Journal of Personality and Social Psychology* 66: pp. 621–32

[21] Voss, C. (2016) '5 Tactics to Win a Negotiation, According to an FBI Agent'. *Time Magazine*, May 25, 2016

[22] Damasio, Antonio R. (2008) [1994]. *Descartes' Error: Emotion, Reason and the Human Brain*. Random House.

[23] Cosgrove, K.P., Mazure, C.M., and Staley, J.K. (2007). 'Evolving knowledge of sex differences in brain structure, function, and chemistry'. *Biol. Psychiatry* 62: pp. 847–855

[24] Ingalhalikar, M., Smith, A., Parker, D., Satterthwaite, T.D., Elliott, M.A., Ruparel, K., Hakonarson, H., Gur, R.E., Gur, R.C. and Verma, R. (2013) 'Sex differences in the structural connectome of the human brain', *Proceedings of the National Academy of Sciences of the United States of America* 111(2)

[25] Kret M.E., De Gelder B. (June 2012). 'A review on sex differences in processing emotional signals', *Neuropsychologia* 50 (7): pp. 1211–21

[26] Cahill L. (June 2006). 'Why sex matters for neuroscience'. *Nature Reviews Neuroscience* 7 (6): pp. 477–84

[27] Cobey, K.D. et al. (2013) 'Sex Differences in Risk Taking Behavior among Dutch Cyclists', *Evolutionary Psychology.*

[28] Ingalhalikar, M., Smith, A., Parker, D., Satterthwaite, T.D., Elliott, M.A., Ruparel, K., Hakonarson, H., Gur, R.E., Gur, R.C. and Verma, R. (2013) 'Sex differences in the structural connectome of the human brain', *Proceedings of the National Academy of Sciences of the United States of America* 111(2)

[29] Keller A, et al. (2012). 'Does the Perception That Stress Affects Health Matter? The Association with Health and Mortality'. Department of Population Health Sciences, University of Wisconsin-Madison; and Jamieson, J., Nock, M. and Mendes, W.B. (2012) 'Mind Over Matter: Reappraising Arousal Improves Cardiovascular and Cognitive Responses to Stress'. *Journal of Experimental Psychology: General*, Vol. 141

[30] Schubert, R., Gysler M., Brown M., Brachinger H.W. (2000) 'Gender Specific Attitudes Towards Risk and Ambiguity: An Experimental Investigation'. Zurich: Center for Economic Research, Swiss Federal Institute of Technology.

[31] Dias-Ferreira, E., Sousa J.C., Melo I., Morgado P., Mesquita A.R., Cerqueira J.J. et al. (2009) 'Chronic stress causes frontostriatal reorganization and affects decision-making'. *Science* 325: pp. 621–625

[32] Dalio, R. (June 2014) 'Company Culture and the Power of Thoughtful Disagreement', *New York Times* Dealbook Conference, available on YouTube

[33] Zaki, J., Schirmer, J., & Mitchell, J.P. (2011). 'Social influence modulates the neural computation of value.' *Psychological Science* 22(7)

[34] Studies show we're four times more likely to remember negative criticism then praise (true even among optimistic people), and that bad feedback is processed more thoroughly than good. Baumeister, Roy F. et al. (2001) 'Bad Is Stronger Than Good', *Review of General Psychology.* Educational Publishing Foundation, 5(4)

ABOUT THE AUTHOR

 Tremaine du Preez is a behavioural economist and international thought leader on a mission to spread the art and science of good decision making. She is a researcher, global faculty member of Duke Corporate Education (a carve-out from Duke's Fuqua School of Business) and consultant to the private and public sectors on organisational decision making.

Since 2008, she has worked across Asia, Africa and Europe, diagnosing decision-making difficulties and building decision-making strategies for individuals and teams up to large multinational organisations. She lectures on behavioural finance, decision science and critical thinking to academic and professional audiences worldwide.

Tremaine is the author and co-author of five books and founder of DECIDE: Decision-Making Consultancy. She has lived in South Africa, Hong Kong and Singapore, and currently lives in London with her family.

www.decideconsultancy.com